SRA
REAL SCIENCE

William C. Kyle, Jr.   Joseph H. Rubinstein   Carolyn J. Vega

A Division of The McGraw-Hill Companies
Columbus, Ohio

Authors

William C. Kyle, Jr.
E. Desmond Lee Family
 Professor of Science Education
University of Missouri – St. Louis
St. Louis, Missouri

Joseph H. Rubinstein
Professor of Education
Coker College
Hartsville, South Carolina

Carolyn J. Vega
Classroom Teacher
Nye Elementary
San Diego Unified School District
San Diego, California

PHOTO CREDITS
Cover Photo: © NASA

SRA/McGraw-Hill
A Division of The McGraw·Hill Companies

Copyright © 2000 by SRA/McGraw-Hill.

All rights reserved. Except as permitted under the United States Copyright Act, no part of this publication may be reproduced or distributed in any form or by any means, or stored in a database or retrieval system, without the prior written permission of the publisher, unless otherwise indicated.

Send all inquiries to
SRA/McGraw-Hill
4400 Easton Commons
Columbus, OH 43219

Printed in the United States of America.

ISBN 0-02-683805-2

8 9 RRW 10 09 08

Content Consultants

Gordon J. Aubrecht II
Professor of Physics
The Ohio State University
  at Marion
Marion, Ohio

William I. Ausich
Professor of Geological
  Sciences
The Ohio State University
Columbus, Ohio

**Linda A. Berne, Ed.D.,
  CHES**
Professor/Health Promotion
The University of
  North Carolina
Charlotte, North Carolina

Robert Burnham
Science Writer
Hales Corners, Wisconsin

Dr. Thomas A. Davies
Texas A&M University
College Station, Texas

Nerma Coats Henderson
Science Teacher
Pickerington Local
  School District
Pickerington, Ohio

Dr. Tom Murphree
Naval Postgraduate School
Monterey, California

Harold Pratt
President, Educational
  Consultants, Inc.
Littleton, Colorado

Mary Jane Roscoe
Teacher/Gifted And
  Talented Program
Columbus, Ohio

Mark A. Seals
Assistant Professor
Alma College
Alma, Michigan

Sidney E. White
Professor Emeritus
  of Geology
The Ohio State University
Columbus, Ohio

Ranae M. Wooley
Molecular Biologist
Riverside, California

Reviewers

Stacey M. Benson
Teacher
Clarksville Montgomery
  County Schools
Clarksville, Tennessee

Mary Coppage
Teacher
Garden Grove Elementary
Winter Haven, Florida

Linda Cramer
Teacher
Huber Ridge Elementary
Westerville, Ohio

John Dodson
Teacher
West Clayton
  Elementary School
Clayton, North Carolina

Cathy A. Flannery
Science Department
  Chairperson/Biology
  Instructor
LaSalle-Peru Township
  High School
LaSalle, Illinois

Cynthia Gardner
Exceptional Children's
  Teacher
Balls Creek Elementary
Conover, North Carolina

Laurie Gipson
Teacher
West Clayton
  Elementary School
Clayton, North Carolina

Judythe M. Hazel
Principal and Science
  Specialist
Evans Elementary
Tempe, Arizona

Melissa E. Hogan
Teacher
Milwaukee Spanish
  Immersion School
Milwaukee, Wisconsin

David Kotkosky
Teacher
Fries Avenue School
Los Angeles, California

Sheryl Kurtin
Curriculum Coordinator, K-5
Sarasota County
  School Board
Sarasota, Florida

Michelle Maresh
Teacher
Yucca Valley
  Elementary School
Yucca Valley, California

Sherry V. Reynolds, Ed.D.
Teacher
Stillwater Public
  School System
Stillwater, Oklahoma

Carol J. Skousen
Teacher
Twin Peaks Elementary
Salt Lake City, Utah

M. Kate Thiry
Teacher
Wright Elementary
Dublin, Ohio

UNIT A

Life Science

Chapter 1   The World of Living Things **A2**

Plants and animals have different structures that help them survive and reproduce.

- **Lesson 1: Plants Inside and Out** **A4**
  - Activity: Moving Water Through Plants A10

- **Lesson 2: Classifying Animals** **A12**
  - Activity: Investigating Vertebrates A18

- **Lesson 3: Adaptations of Plants and Animals** **A20**
  - Activity: Animal Behavior A26

Chapter Review **A28**

Chapter 2   Organisms Live and Grow **A30**

Organisms need energy and matter to live and grow.

- **Lesson 1: Cycles of Life** A32
  - Activity: Comparing Animal Life Cycles A38

- **Lesson 2: Carbon and Water Cycles** A40
  - Activity: Observing Transpiration A46

- **Lesson 3: Energy Flow** A48
  - Activity: Food Webs A54

Chapter Review **A56**

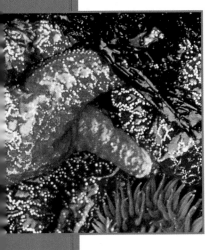

Chapter 3 Ecosystems **A58**

An ecosystem represents the interactions between living and nonliving things.

Lesson 1: Characteristics of Ecosystems .. A60
Activity: Influencing an Ecosystem A66

Lesson 2: Ocean Ecosystems A68
Activity: Decomposers A74

Lesson 3: Changing Ecosystems A76
Activity: Adding to Ecosystems A82

Chapter Review **A84**

Unit Review **A86**

UNIT B

Earth Science

Chapter 1 The Solar System B2

Gravity is a force that exists between objects.

Lesson 1: Gravity and the Sun B4
Activity: Modeling the Solar System B12

Lesson 2: The Moon and Earth B14
Activity: Gravitational Pull B20

Lesson 3: Space Exploration B22
Activity: Making a Balloon Rocket B28

Chapter Review B30

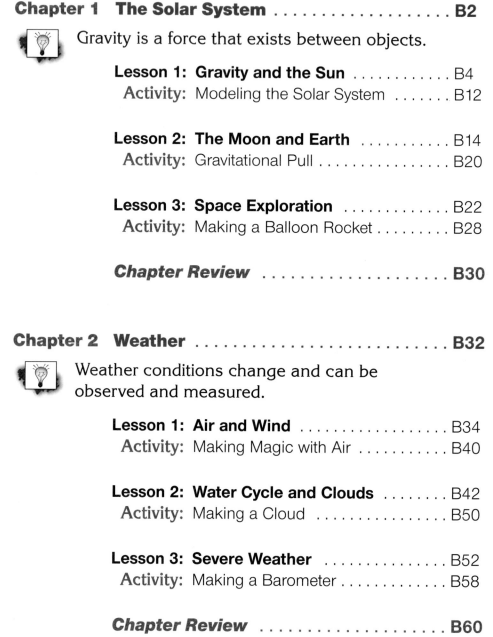

Chapter 2 Weather B32

Weather conditions change and can be observed and measured.

Lesson 1: Air and Wind B34
Activity: Making Magic with Air B40

Lesson 2: Water Cycle and Clouds B42
Activity: Making a Cloud B50

Lesson 3: Severe Weather B52
Activity: Making a Barometer B58

Chapter Review B60

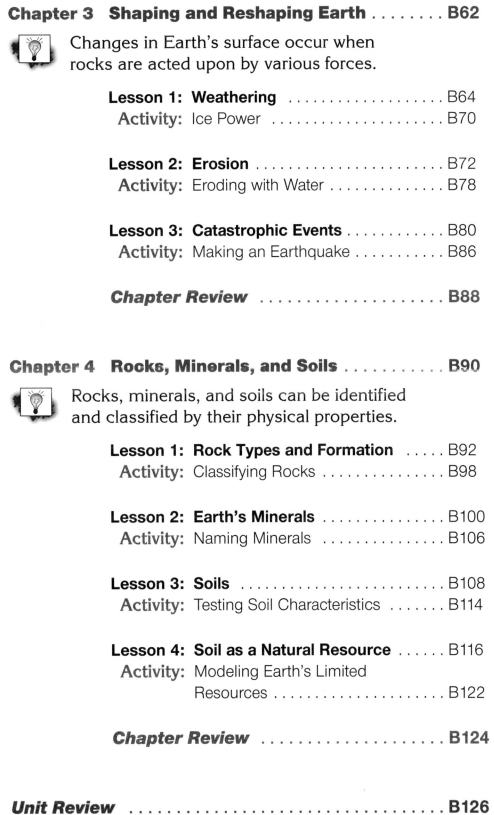

Chapter 3 Shaping and Reshaping Earth **B62**

Changes in Earth's surface occur when rocks are acted upon by various forces.

Lesson 1: Weathering B64
Activity: Ice Power B70

Lesson 2: Erosion B72
Activity: Eroding with Water B78

Lesson 3: Catastrophic Events B80
Activity: Making an Earthquake B86

Chapter Review **B88**

Chapter 4 Rocks, Minerals, and Soils **B90**

Rocks, minerals, and soils can be identified and classified by their physical properties.

Lesson 1: Rock Types and Formation B92
Activity: Classifying Rocks B98

Lesson 2: Earth's Minerals B100
Activity: Naming Minerals B106

Lesson 3: Soils B108
Activity: Testing Soil Characteristics B114

Lesson 4: Soil as a Natural Resource B116
Activity: Modeling Earth's Limited
Resources B122

Chapter Review **B124**

Unit Review **B126**

UNIT C

Physical Science

Chapter 1 Static Electricity and Magnets C2

💡 Electricity and magnets can exert force.

Lesson 1: Static Electricity C4
Activity: Opposites Attract, Likes Repel C10

Lesson 2: Magnetism C12
Activity: Magnetic Fields C18

Chapter Review C20

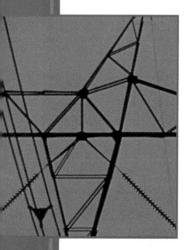

Chapter 2 Energy Pathways C22

💡 Electric current moves through different circuits.

Lesson 1: Electric Circuits C24
Activity: Make a Bulb Light Up C32

Lesson 2: Series and Parallel Circuits C34
Activity: Making Circuits C40

Lesson 3: Electromagnetism C42
Activity: Turning a Magnetic Field
On and Off C48

Chapter Review C50

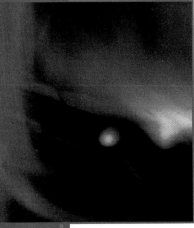

Chapter 3 Heat **C52**

Heat is produced and transferred in many ways.

    Lesson 1: Heat Production C54
      Activity: Producing Heat C60

    Lesson 2: Heat Transfer C62
      Activity: Observing Heat Transfer
              by Conduction C68

    Lesson 3: Using Heat C70
      Activity: Using Heat to Do Work C76

    Chapter Review **C78**

Chapter 4 Matter, Motion, and Machines **C80**

Because matter has mass, it takes force to move it.

    Lesson 1: Matter C82
      Activity: Comparing Density C88

    Lesson 2: Motion C90
      Activity: Observing Inertia C96

    Lesson 3: Simple Machines C98
      Activity: Using Simple Machines C106

    Chapter Review **C108**

Unit Review **C110**

UNIT D

Health Science

Chapter 1 Skin **D2**

Skin protects the body and controls the body's temperature.

Lesson 1: Properties of Skin D4
Activity: The Sense of Touch D10

Lesson 2: Skin Controls Body Temperature D12
Activity: Facial Wrinkles D18

Chapter Review D20

Chapter 2 Chemical Substances **D22**

Chemical substances affect the body in many ways.

Lesson 1: Chemical Substances Cause Changes D24
Activity: Educating Others D30

Lesson 2: Positive Effects of Chemical Substances D32
Activity: Reading the Label D38

Lesson 3: Abuse of Chemical Substances D40
Activity: Tracing the Effects of Chemical Substances D46

Chapter Review D48

Chapter 3: Nutrition **D50**

💡 Nutrients help the body to grow healthy and strong.

Lesson 1: Carbohydrates, Fats, and Proteins D52
Activity: Finding Fats and Starch in Foods D58

Lesson 2: Water, Vitamins, and Minerals ... D60
Activity: Reading a Food Label D66

Chapter Review D68

Unit Review D70

Reference R1

Science Process Skills

Understanding and using scientific process skills is a very important part of learning in science. Successful scientists use these skills in their work. These skills help them with research and discovering new things.

Using these skills will help you to discover more about the world around you. You will have many opportunities to use these skills as you do each activity in the book. As you read, think about how you already use some of these skills every day. Did you have any idea that you were such a scientist?

Observing

Use any of the five senses (seeing, hearing, tasting, smelling, or touching) to learn about objects or events that happen around you.

> **Looking** at objects with the help of a microscope is one way to observe.

Communicating

Express thoughts, ideas, and information to others. Several methods of communication are used in science—speaking, writing, drawing graphs or charts, making models or diagrams, using numbers, and even body language.

> **Making a graph** to show the rate of growth of a plant over time is communicating.

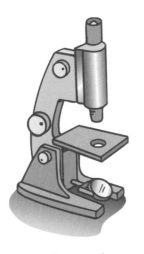

Classifying

Organize or sort objects, events, and things that happen around you into categories or groups. The classified objects should all be alike in some way.

> **Sorting** students in the room into groups according to hair color is classifying.

Using Numbers

Use math skills to help understand and study the world around you. These skills include ordering, counting, adding, subtracting, multiplying, and dividing.

Comparing the temperatures of different locations around your home is using numbers.

Measuring

Use standard measures of time, distance, length, area, mass, volume, and temperature to compare objects or events. Measuring also includes estimating and using standard measurement tools to find reasonable answers.

Using a meterstick to find out how far you can jump is measuring.

Constructing Models

Draw pictures or build models to help tell about thoughts or ideas or to show how things happen.

Drawing the various undersea formations on the ocean floor is constructing a model.

Inferring

Use observations and what you already know to reach a conclusion about why something happened. Inferring is an attempt to explain a set of observations. Inferring is not the same as guessing because you must observe something before you can make an inference.

Imagine that you put a lettuce leaf in your pet turtle's aquarium. If the lettuce is gone the next day, then you can **infer** that the turtle ate the lettuce.

Predicting

Use earlier observations and inferences to forecast the outcome of an event or experiment. A prediction is something that you expect to happen in the future.

Stating how long it will take for an ice cube to melt if it is placed in sunlight is **predicting.**

Interpreting Data

Identify patterns or explain the meaning of information that has been collected from observations or experiments. Interpreting data is an important step in drawing conclusions.

You interpret data when you **study** daily weather tables and **conclude** that cities along the coast receive more rainfall than cities in the desert.

Identifying and Controlling Variables

Identify anything that may change the results of an experiment. Change one variable to see how it affects what you are studying. Controlling variables is an important skill in designing investigations.

You can **control** the amount of light plant leaves receive. Covering some of the leaves on a plant with foil allows you to compare how plant leaves react to light.

Hypothesizing

Make a statement that gives a possible explanation of how or why something happens. A hypothesis helps a scientist design an investigation. A hypothesis also helps a scientist identify what data to collect.

Saying that bean seeds germinate faster in warm areas than cold areas is a hypothesis. You can **test** this hypothesis by germinating bean seeds at room temperature and in the refrigerator.

Defining Operationally

An operational definition tells what is observed and how it functions.

Saying the skull is a bone that surrounds the brain and is connected to the backbone is an operational definition.

Designing Investigations

Plan investigations to gather data that will support or not support a hypothesis. The design of the investigation determines which variable will be changed, how it will be changed, and the conditions under which the investigation will be carried out.

You can **design an investigation** to determine how sunlight affects plants. Place one plant in the sunlight and an identical plant in a closet. This will allow you to control the variable of sunlight.

Experimenting

Carry out the investigation you designed to get information about relationships between objects, events, and things around you.

Experimenting pulls together all of the other process skills.

UNIT A

Life Science

Chapter 1 **The World of Living Things** **A2**
    Lesson 1: Plants Inside and Out A4
        Activity: Moving Water Through Plants A10
    Lesson 2: Classifying Animals A12
        Activity: Investigating Vertebrates A18
    **Lesson 3: Adaptations of Plants
and Animals** A20
        Activity: Animal Behavior A26
    Chapter Review **A28**

Chapter 2 **Organisms Live and Grow** **A30**
    Lesson 1: Cycles of Life A32
        Activity: Comparing Animal Life Cycles A38
    Lesson 2: Carbon and Water Cycles A40
        Activity: Observing Transpiration A46
    Lesson 3: Energy Flow A48
        Activity: Food Webs A54
    Chapter Review **A56**

Chapter 3 **Ecosystems** **A58**
    Lesson 1: Characteristics of Ecosystems .. A60
        Activity: Influencing an Ecosystem A66
    Lesson 2: Ocean Ecosystems A68
        Activity: Decomposers A74
    Lesson 3: Changing Ecosystems A76
        Activity: Adding to Ecosystems A82
    Chapter Review **A84**

Unit Review **A86**

CHAPTER 1

The World of Living

Life has been on Earth for longer than we can even imagine. Scientists think the first signs of life on Earth appeared over 3.5 billion years ago! Since then, all living organisms that have been on Earth have carried on what we call life functions. They have grown, they have changed, and they have reproduced.

What is the smallest living organism you have ever seen? A tiny insect on a leaf? Tiny green plants on the forest floor? Actually, the most abundant types of organisms on Earth can be seen only with a microscope. The largest living organism, the blue whale, can be as big as a house. Even though living organisms come in many shapes and sizes, they have some basic traits in common.

The Big IDEA

Plants and animals have different structures that help them survive and reproduce.

Things

CHAPTER SCIENCE INVESTIGATION

Observe how a plant species has adapted. Find out how in your *Activity Journal.*

Lesson 1

Plants Inside and Out

Find Out
- What functions the different plant structures perform
- How plants reproduce
- How plant structures help scientists classify and study plants

Vocabulary
photosynthesis
chlorophyll
spore
vascular plants
nonvascular plants

The Big QUESTION
How are plants classified?

A cactus, a fern, a daisy, a pine tree, a tomato plant. Could these plants be more different from each other? Well . . . yes! And . . . no! One plant's structures can look very different from another plant's structures, but they perform the same tasks.

Plant Structures

In any organism, including plants, cells are very tiny parts that perform special functions. Most cells are so small that you need a microscope to see them. But when cells join together, they build tissues and organs that perform functions to keep the organism alive. Most plants have leaves, stems, and roots. These plant organs work together to perform the plant's life functions.

One plant function is to produce sugar for food so the plant can stay alive. Green plants use sunlight, nutrients in the soil, water, and a gas called carbon dioxide to make the sugars they need through the process of **photosynthesis** (fōt′ ə sin′ thə sis). Photosynthesis takes place in some cells of the plant's leaves where the green substance **chlorophyll** (klor′ ə fil′) is found.

Plants take carbon dioxide in through openings on the leaf. When sunlight reaches the leaves, the chlorophyll inside the leaf uses the sun's energy to make sugars, which are stored in the plant. In this process, water and carbon dioxide are turned into oxygen and sugar. The plant uses the sugar for food and gives off the oxygen into the air.

Plants don't eat food the way you do. They make their own food inside the leaves.

Leaf

Stem The stem moves nutrients and water from the roots to the leaves and sugars from the leaves to the roots.

O_2

CO_2

Chloroplasts

Roots

Chloroplasts are the tiny structures that contain chlorophyll. This is where photosynthesis happens.

Roots absorb water and nutrients from the soil and anchor the plant in the ground. They carry nutrients and water from the soil to the rest of the plant.

A5

Plant Reproduction

Flowers

Another organ found in many plants is the flower. The flower is the part of the plant where seeds form. Before seeds can form, pollination must take place. Pollination happens inside the flower when pollen grains are moved onto the pistil. Then a tube grows from each pollen grain downward into the plant's ovary. There, fertilized cells develop into embryos and seeds are formed. Both pollination and fertilization must take place to produce seeds that will grow into new plants. Wind or insects such as bees are two ways that pollen can be moved onto the flower's pistil.

Seeds

Flowers are adapted to produce seeds. Some flowers, such as peaches and cherries, produce only one seed in each fruit. Others, such as apples, oranges, and beans, produce more seeds. Still other flowers produce many seeds at the same time. For example, milkweed and dandelion plants send out many seeds when they burst open. In this way the wind can blow the seeds great distances where they take root and start new plants.

The stamen produces a powdery material called pollen. Pollen grains will fertilize the cells in the flower's ovary.

The petals are the parts of the flower that surround the inside parts. They are often brightly colored and attract insects that can pollinate the plant.

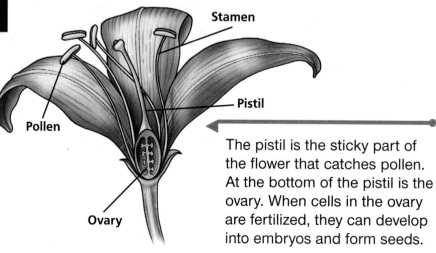

The pistil is the sticky part of the flower that catches pollen. At the bottom of the pistil is the ovary. When cells in the ovary are fertilized, they can develop into embryos and form seeds.

Other Ways to Reproduce

Not all seed-producing plants have flowers. Pines, firs, and spruce trees are plants that make seeds inside of cones. Other plants that make seeds without flowers are ginkgoes and cycads.

Still other plants do not produce seeds at all. Instead, these plants reproduce by forming tiny cells called spores. A **spore** (spōr) is a cell that develops into a new organism. When spores fall on moist ground, they grow into new plants. Mosses and ferns are common spore-producing plants. These plants grow in clusters close to the ground in wet or moist areas.

(Top) Seeds are made inside of cones. (Bottom) Ferns produce spores instead of seeds.

Ginkgo trees produce seeds but no flowers.

Classifying Plants

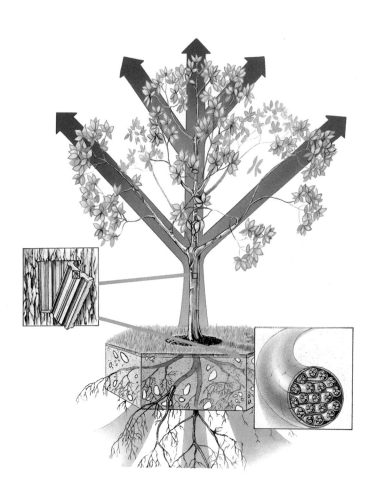

If a vascular plant needs water, all you have to do is water the roots. The plant's tubelike cells move water from the ground to the stems and leaves where it's needed.

What if you found a plant no one had ever seen before? How could you compare your discovery to other plants? You might look at its shape. Does it grow tall or very close to the ground? You might look at its organs of reproduction. Does it have flowers? Does it have seeds? Does it have cones? You might look at its leaves. Are they needle-like, broad and flat, or small and pointy?

Scientists group plants first by finding out how they take water up from their roots to their leaves. Some plants have tiny tubes that move water and nutrients inside the plant. If you look at a lettuce leaf, you can see these tubes. Other plants, such as mosses, do not have these tubes. Whether or not a plant has tubelike structures is a good first step to distinguish between plants.

Plants that move water from the roots to the stems and leaves are called **vascular** (vas′ kyə lər) **plants.** Vascular plants include flowering plants, other seed-producing plants, and ferns. Vascular plants carry water to other parts of the plant in a way similar to the way plumbing carries water throughout your house. You don't store water in the kitchen or the bathroom. Pipes carry the water from beneath the ground, up through the walls of your house, to the rooms where it is needed. In the same way, tubes carry water and nutrients from the soil to the parts of the plant where they are needed.

Plants without tube-like parts are called **nonvascular** (non vas′ kyə lər) **plants.** Nonvascular plants don't have roots. These plants absorb water through their surfaces. Mosses and worts are nonvascular plants. Because they must absorb water to survive, these plants grow only in moist areas.

Another way to classify plants is to look at the way they reproduce. Seed plants include the daisies in your garden, the shrubs around your house, the pine trees in the park, and the vegetables that we eat. All of these plants produce seeds but not all of them flower.

From the large group of flowering plants, still more groups are classified according to other similarities and differences, such as what their leaves look like or how their flowers are formed. Today scientists use 12 main groups for classifying over 260,000 species of plants! And scientists are still finding more.

Ferns are vascular plants. Mosses are nonvascular plants. Why can ferns grow tall while mosses grow close to the ground?

What types of plants can you find in this photo?

CHECKPOINT

1. What functions do the roots, stems, leaves, and flowers have?
2. What are two ways plants can reproduce?
3. What are two main things scientists look at when they begin to study a new plant?

 How are plants classified?

ACTIVITY

Moving Water Through Plants

Find Out
Do this activity to learn how tubelike structures in plants conduct water from the roots to the leaves.

Process Skills
Communicating
Measuring
Controlling Variables
Observing
Inferring

What You Need

tape, meter tape, pencil, paper towel, hand lens, lettuce leaf, moss, scissors, 0.5-mm bore capillary tube, two plastic jars, clock with a second hand, container of water colored with food coloring, Activity Journal

What to Do

1. Cut a piece of paper toweling 2 cm wide. Tape one end of the paper towel strip to the middle of a pencil. Make sure the strip is long enough to go from the top of the plastic jar to the bottom of the jar.

2. Use the meter tape and mark each plastic jar 3 mm up from the bottom. Fill each jar with colored water to reach the 3-mm mark.

A10

3. Stand the glass tube in one jar, hold it in place, and **time** how long it takes for the water to reach its highest level.

   Safety! *Glass tubes can break, so handle with care.*

4. Place the pencil across the mouth of the other jar, with the paper toweling hanging down into the jar.
5. **Observe** and **time** how long it takes for the water in the paper towel to rise to the same height as the water in the glass tube.
6. **Record** your observations.
7. **Examine** the moss and the lettuce with a hand lens. **Decide** which plant is like the paper towel and which is like the glass tube.

CONCLUSIONS

1. How long did it take for the water to rise in the glass tube?
2. How long did it take for the water to rise in the paper towel?
3. Which plant is like the paper towel and which one is like the glass tube?

ASKING NEW QUESTIONS

1. Why do you think mosses are small and grow close to the ground?
2. Do you think trees absorb water like the paper towel or like the glass tube? Why?

SCIENTIFIC METHODS SELF CHECK

✔ Did I **record** what happened to the water in each jar?

✔ Did I **measure** the same amount of water for each jar?

✔ Did I **compare** how the moss and the lettuce are like the paper towel and the glass tube?

LESSON 2

Classifying Animals

Find Out
- What kinds of animals don't have backbones
- How animals with backbones can differ
- What are other ways animals can be different

Vocabulary
invertebrates
vertebrates
cartilage
warm-blooded
cold-blooded
symmetry

The Big QUESTION
What makes animals different from one another?

*R*each around and feel the middle of your back. What you feel is your backbone. What animals can you think of that have backbones? Would you be surprised to find out that 95 percent of animals on Earth do not have backbones?

Invertebrates

Scientists estimate that there are more than one million kinds of animals on Earth. There are lots of ways you might put these animals into groups. There are big animals, small animals, tame ones, and wild ones. Some animals live in the water. Some animals fly in the air. Scientists classify animals in a different way than they classify plants. Just like plants, all animals are made of cells. These cells combine into different kinds of tissues, organs, and systems.

One way to classify animals is to see if they have a backbone. Animals without a backbone are called **invertebrates** (in vûr′ tə brāts). Many invertebrates, such as worms, have soft bodies. The simplest worms are called flatworms. They have long, flat bodies and live in water or inside of other organisms. A tapeworm is a flatworm that attaches itself to the intestine of another animal and absorbs the animal's food. Earthworms and leeches are segmented (seg′ mənt əd) worms. They have bodies that are divided into rings or segments. Segmented worms have systems for digesting food, getting rid of wastes, and moving blood.

Another group of soft-bodied invertebrates that may have hard shells are called mollusks. You might have eaten mollusks. Clams, scallops, oysters, and snails are in this group. Some, like the squid and the slug, have internal shells that you don't see. Octopuses have no shell.

Insects, spiders, and centipedes belong to the group called arthropods (är′ thrə pods). Crabs, crayfish, shrimp, and lobsters are also arthropods. They have an outer skeleton and legs that bend at joints. Arthropods also have bodies that are divided into sections. The outer skeleton, or exoskeleton, is shed several times as the arthropod grows to adulthood.

Different types of invertebrates include an earthworm, a tarantula, and flame scallops.

Photos not to scale

Vertebrates

Animals with backbones are called **vertebrates.** Because you have a backbone, you have something in common with other vertebrates. Their skeletons are on the inside of their bodies, and are made of bone or cartilage. **Cartilage** (kär′ tə lej) is a tough, flexible material that helps to support and shape body parts. The end of your nose and the tops of your ears are made of cartilage.

Warm-Blooded Vertebrates

Body temperature is one way to group vertebrates. Of all the vertebrates, only birds and mammals have the ability to control their body temperatures. They have the same body temperatures when it is cold outside as they do when it is hot outside. Animals that can control their body temperatures are called **warm-blooded** animals. You are a warm-blooded animal.

Birds are also warm-blooded vertebrates, and most are able to fly. Just as skin and hair protect your body, feathers protect a bird's body and help control its body temperature. Some birds, such as penguins, are not able to fly. Instead, they use their wings as flippers to help them swim.

Mammals include animals such as goats, wolves, whales, bats, and humans. Mammals are warm blooded with hair or fur on part of their bodies. Female mammals produce milk to feed their young. Some mammals, like whales, dolphins, seals, and manatees, live in the sea. Mammals can be adapted to hot or very cold climates.

A turkey is a vertebrate because it has a skeleton on the inside of its body.

A koala is a mammal.

Photos not to scale

Cold-blooded vertebrates include a raccoon butterfly fish, an American toad, and a striped whipsnake.

Cold-Blooded Vertebrates

Some vertebrates are classified as **cold-blooded** vertebrates because their body temperatures stay the same as the temperature around them. Fish are cold-blooded vertebrates. They need oxygen to live just as you do, but they get the oxygen they need from water. Water flows into a fish's mouth and out past its gills, where the oxygen is collected. Most fish are covered with scales for protection.

Another group of vertebrates is called amphibians. Frogs, salamanders, and toads are amphibians. The word amphibian means "having two lives." An amphibian can live part of its life in the water and part on land.

Another group of vertebrates includes reptiles. Crocodiles, alligators, snakes, and lizards are in this group. Reptiles are cold-blooded animals that have dry, scaly skin. Most reptiles live on land, breathe with lungs, and lay eggs with tough shells. Except for snakes and some lizards, most reptiles have two pairs of legs.

Other Ways Animals Can Differ

What Animals Eat

Animals can be grouped by what they eat. Carnivores (kär′ nə vorz) are animals that eat mostly meat. Sharks, eagles, and lions are carnivores. The teeth on most carnivores are long and sharp for biting and tearing meat. Birds that are carnivores have sharp, curved beaks to rip and tear.

Animals like cows and kangaroos that feed mainly on plants are called herbivores (hûr′ bə vorz). Compare the teeth of a herbivore with those of a carnivore. Cows and sheep eat grass, and they have flat teeth to grind up the plants. Some birds that are herbivores eat only seeds. Their beaks are short and cone-shaped to crush and grind up the seeds.

Some animals eat both meat and plants. They are called omnivores (om′ ni vorz). Baboons and opossums are omnivores. What are your eating habits? Are you an herbivore, a carnivore, or an omnivore?

A red fox and a shark are carnivores. Their teeth are long and sharp.

A European goldfinch has a short, cone-shaped beak to crush seeds.

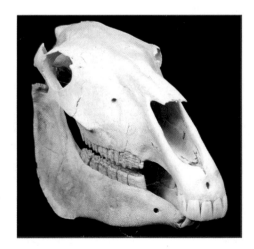

A horse is an herbivore. Its teeth are flat to grind up plants.

How Bodies Are Arranged

In addition to what animals eat, scientists classify animals by how their body parts are arranged. This is called the animal's **symmetry** (sim′ ə trē). Some animals have two halves that look very similar. If you draw an imaginary line down a cat's body, you will see that each side has about the same number of parts—one eye, one ear, two legs, etc. This is called bilateral (bī lat′ ə rel) symmetry. Most vertebrates are shaped this way.

A Promethea moth has bilateral symmetry.

Some animals, however, are arranged around a central point, like the spokes on a wheel. They do not have a distinct right side and left side. This type of body arrangement is called radial (rād′ ē əl) symmetry. Sea anemones and sea stars are shaped this way.

A sea anemone has radial symmetry.

There are also some animals that don't have a balanced shape. They cannot be divided into equal parts. This is called asymmetry (ā sim′ ə trē). Many sponges are shaped like this. What kind of symmetry does your body have?

A sponge has asymmetry.

CHECKPOINT

1. What kinds of animals don't have backbones?
2. What is the difference between warm-blooded and cold-blooded vertebrates?
3. Describe other ways animals can be classified.// What makes animals different from one another?

A17

ACTIVITY

Investigating Vertebrates

Find Out
Do this activity to learn how vertebrates are alike and different.

Process Skills
Observing
Communicating
Measuring
Using Numbers
Inferring
Classifying
Defining Operationally

What You Need

metric ruler

Activity Journal

What To Do

1. **Observe** each animal photograph carefully. Note the size of the animals in the photographs by reading the scale provided for each. **Record** the information for the steps that follow in a data table.

2. Identify the animal. **Measure** its length in the photo in centimeters. Use the scale to **calculate** how long or tall the animal is.

3. Identify the color of the animal.

70 cm

15 cm

4. **Observe** the animal body parts. How are they arranged? **Describe** the type of outer covering the animal has.
5. **Infer** how you think the animal moves. **Infer** what type of environment the animal needs.

CONCLUSIONS

1. Based upon your observations, how could you **classify** the different animals?
2. What body parts help each of the animals move?
3. What is the smallest animal? The largest?

ASKING NEW QUESTIONS

1. What are the major differences you observed among fish, amphibians, and reptiles?
2. What are the major differences you observed between birds and mammals?

SCIENTIFIC METHODS SELF CHECK

✔ Did I **observe** the animals carefully?

✔ Did I **record** my information in a data table?

✔ Did I **measure** the length of the animals using centimeters?

✔ Did I **infer** the way the animals move and the type of environments the animals need?

✔ Did I **classify** the animals and define how they differ?

Lesson 3

Adaptations of Plants and Animals

Find Out
- What a species is
- How animal and plant species adapt to survive
- How animal behaviors can be inherited or learned

Vocabulary
species
inherited traits
adaptation
instincts
learned behavior

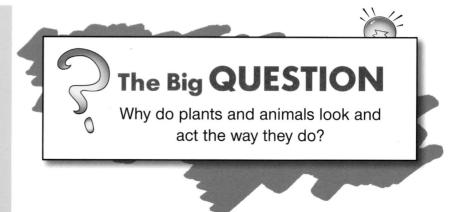

The Big QUESTION

Why do plants and animals look and act the way they do?

If an elephant wants to get some leaves from a tree, what do you think it might do? How would a monkey go after those same leaves? An insect? Animals have different ways of getting the things they need. So do plants.

Species of Organisms

There are different groupings of organisms. A **species** (spē′ sēz) is a group of organisms that usually reproduces only with other organisms of the same species. For example, dogs can reproduce with other dogs. Even though dalmatians look quite different from Irish setters, they are still able to reproduce and have puppies. Dalmatians and Irish setters belong to the same species. Dogs and cats are in different groupings, including different species. They cannot reproduce and have young.

Just as there are different species of animals, there are also different species of plants. A cherry tree can pollinate the same kind of cherry tree and produce seeds. But a sugar maple tree cannot pollinate a silver maple tree and produce seeds. The sugar maple belongs to a different species than the silver maple tree does.

When a species reproduces, it passes along the characteristics of that species. Those characteristics that are passed along to their young are called **inherited traits.** Inherited traits can be important to help a species survive.

The leaves from a white maple, a sugar maple, and a red maple may look similar but they are from different species.

A dalmatian and an Irish setter look quite different, but they are the same species.

Structural Adaptations

An **adaptation** is anything that helps a species live and reproduce in its environment. Different structural adaptations in plant and animal species that have occurred over time help them to survive. Species have adapted to many different environments on Earth because of their structural adaptations.

The leaves on a spruce tree are like tiny needles. Their long, narrow shape allows the tree to withstand severe winds in cold climates. Their hard surface also prevents them from drying out in windy climates.

The elephant ear plant lives on the forest floor under the shade of taller plants. Its huge leaves collect as much sunlight as possible. It also has a purple layer on the underside of the leaves that reflects some of the light that passes through the leaf.

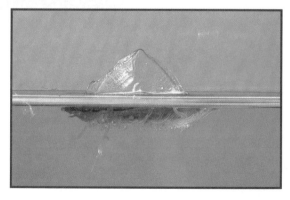

The by-the-wind-sailor does not swim. Instead, its gas-filled sail allows it to float along in the water.

Structural adaptations can also allow an organism to protect itself from predators. An adaptation that allows an organism to blend into its environment is called *camouflage*. The living stone plant in Africa is a camouflaged plant. The small round leaves blend in with the rocky ground so the plant is not eaten.

Animals can also mimic other organisms as a way to protect themselves. The hawkmoth caterpillar can imitate a poisonous snake. When a predator comes near, the caterpillar can turn its back end around to look like the face of a poisonous snake. The spots on the caterpillar look like the eyes of the snake.

Baobab trees have thick trunks that can store water for the dry season. They also have tough bark to protect them against fire.

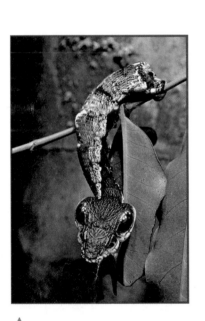

The hawkmoth caterpillar can mimic a poisonous snake. Spots on the caterpillar's back look like the eyes of a snake.

The Goliath heron has widespread toes which stop the heron from sinking into the muddy soil.

The living stone plant is camouflaged and blends into the rocky ground.

Animal Behavior

The ways animals act can also be adaptations. Everything an animal does is part of its behavior. Animals can use a variety of actions to protect themselves. Killdeer are birds that can pretend to be hurt in order to protect themselves and their young. When a predator approaches the nest, one of the parents limps away as if its wings were broken. After the enemy is lured away from the young killdeer, the parent flies away to safety.

Instincts are behaviors that animals are born with. They are inherited from their parents just like structural adaptations. For example, bees are born knowing how to make honey. A spider knows right away how to spin a web to catch its food. A kitten turns to its mother for milk when it is born. Many birds know how to make nests. An animal knows how to find food, and to mate to produce offspring.

Some animals migrate, or move to better conditions, by instinct. Some animals migrate to warmer weather for food and a safe place to raise their young. These monarch butterflies will travel from Mexico all the way north to Canada. How do they know where to go? Instinct.

Other animal behaviors are not inherited from their parents. If you have ever tried to train a pet, you know that not all behaviors are instinct. **Learned behavior** is an action that is changed by experience. A particular rat can learn how to follow a certain path in a maze. A bird called a macaw can learn to say words. You can also learn how to do many things. However, learned behaviors are not passed along to offspring. A macaw may learn to say hello, but that skill will not be inherited by its young. Learned behaviors must be experienced by every individual in the species.

All living organisms are what they are and the way they are because of the combination of what they inherit from their parents and what they learn from their environment.

Animals can learn many different kinds of behaviors. Some behaviors are helpful for humans. Other behaviors can be harmful for humans or for the animals themselves. When a bear learns that food can be found near humans, both the humans and the bear can be in danger.

CHECKPOINT

1. What is a species?
2. What is one example of an animal and one example of a plant adaptation?
3. Give two examples each of inherited and learned behaviors.

 Why do plants and animals look and act the way they do?

ACTIVITY

Animal Behavior

Find Out
Do this activity to learn how earthworms behave when placed in a new environment.

Process Skills
Observing
Communicating
Controlling Variables

What You Need

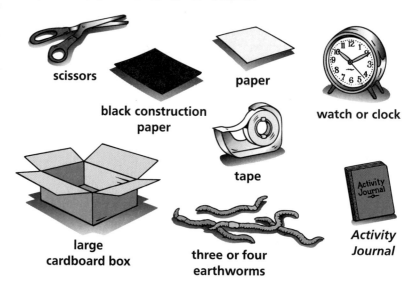

scissors

black construction paper

paper

tape

watch or clock

large cardboard box

three or four earthworms

Activity Journal

What to Do

1. Tape a sheet of paper on a table or desk. Cut the black construction paper in half.
2. Fold the black paper to look like a tent and place it on one end of the paper.
3. Place three or four earthworms on the other end of the paper.
4. **Observe** the movement of the earthworms for three minutes and then **record** your observations.

Safety! *Wash your hands after handling the earthworms.*

5. Leave the worms for one or two hours. After that time, observe and record where the worms are.

6. Now repeat Steps 3 and 4, but this time place the cardboard box over the entire sheet of paper. After one hour, pick up the box and observe where the earthworms moved.

CONCLUSIONS

1. Where did the earthworms move after one or two hours?
2. Did all of the earthworms move to the same location?
3. Compare the positions of the earthworms under the box and on the paper after one hour.

ASKING NEW QUESTIONS

1. Why do you think the earthworms reacted the way they did?
2. Can you explain how this behavior might be an example of instinct?

SCIENTIFIC METHODS SELF CHECK

✔ Did I **observe** several earthworms for the required amount of time?

✔ Did I **record** my observations?

✔ Did I **observe** each earthworm for the same length of time?

✔ Did I **compare** the positions of the earthworms after one hour?

Review

Reviewing Vocabulary And Concepts

Match the correct word with its definition.

Set A

1. process in which green plants use sunlight, water, nutrients, and carbon dioxide to create sugar for food
2. plants that move water from the roots to the stems and leaves
3. plants that absorb water only through their surfaces
4. animals whose body temperature is the same as the temperature around them
5. animals that maintain a constant body temperature

a. warm-blooded animals
b. vascular plants
c. photosynthesis
d. cold-blooded animals
e. nonvascular plants

Set B

1. an animal that eats both plants and animals
2. a group of organisms that usually only reproduce within their group
3. anything that helps a species live within its environment
4. behaviors that animals are born with
5. an action that is changed by one's experience

a. species
b. omnivore
c. learned behavior
d. adaptation
e. instincts

Review

Understanding What You Learned
1. How does sunlight help plants to survive?
2. What is chlorophyll and where is it found in plants?
3. What function do seeds and spores perform in plants?
4. Name one characteristic that helps people classify plants and one characteristic that helps people classify animals.
5. What are some behaviors or adaptations that help plants and animals to survive in their environment?

Applying What You Learned
1. Describe the functions of the different parts of a plant's roots, stems, leaves, and flowers. You may want to use an illustration to help you explain.
2. What are some ways flowers can be pollinated?
3. How could you use a mirror to test different types of body symmetry?
4. Name an example of an instinct and a learned behavior in an animal and in a human being.

5. Why do plants and animals have different structures?

For Your Portfolio

Pick an area on Earth and draw an imaginary plant and animal that could survive there. Include how the plant would reproduce and what adaptations the animal would have. Possible areas might be the top of the Rocky Mountains, Antarctica, the bottom of a rain forest, the middle of a desert, or the middle of the Pacific Ocean.

CHAPTER 2

Organisms

Life on Earth is a struggle. Without basic requirements organisms die. To grow and reproduce, all organisms need space and water. Most importantly, organisms need energy. Directly or indirectly, they all get that energy from the sun.

Organisms can be divided into two basic categories: those that produce food and those that consume it. The food that links organisms into a food chain provides them with energy. Energy can be transferred from one organism to another within a system. Water and other substances can also move through systems. These cycles allow for life to continue on Earth.

The Big IDEA

Organisms need energy and matter to live and grow.

Live and Grow

CHAPTER SCIENCE INVESTIGATION

Learn how brine shrimp grow and develop. Find out how in your *Activity Journal.*

A31

Lesson 1

Cycles of Life

Find Out
- How plants change during their life cycles
- How animals change during their life cycles
- Why we should know about life cycles

Vocabulary
life cycle
germination
metamorphosis
larva
pupa

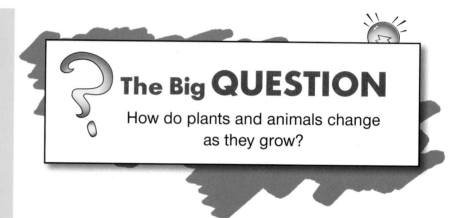

The Big QUESTION
How do plants and animals change as they grow?

Sometimes things happen over and over in the same way. For example, the sun rises every morning, sets again at night, and then rises again the next day. Winter is followed by spring, then summer, then fall, and then winter again. These are parts of Earth's natural cycles. Natural cycles are also found when we study the lives of organisms.

Plant Life Cycles

All living things go through a cycle of four basic stages in life: a beginning, a growth period, reproduction, and death. These stages make up what we call the **life cycle.** Organisms start as single cells. Remember how flowering plants are pollinated? When a pollen grain fertilizes a cell in a flower, an embryo grows in the seed. Under the right conditions, the seed will sprout and grow. This plant will produce flowers that will produce more seeds if they are pollinated. If the life cycle continues, the species can continue to reproduce.

A32

Some plants, like tomatoes, grow, reproduce, and die all in one year. Other plants, like trees or shrubs, live for more than one year. They continue to grow and reproduce regularly. Still other plants, such as carrots, produce leaves and food one year, and then reproduce and die the next year. Different kinds of life cycles help plant species to survive in different environments.

3–4. Reproduction and Death
After reproducing, some plants will die. Others will live longer and reproduce again. Although the final stage for a mature plant is death, that doesn't mean the end of things. The seeds and new plants that a parent plant produces start the life cycle all over again.

2. Growth
The young plant that pops through the ground is called a seedling. Without the right amount of light, water, food, and air, the plant cannot mature and reproduce.

1. *Germination*
With the right conditions, a root, stem, and leaf will grow out of a seed. This sprouting of new plant structures is called **germination.**

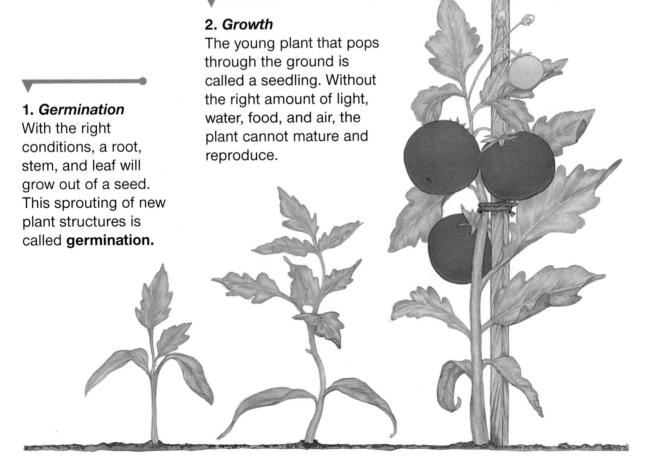

Animal Life Cycles

Animals also go through different stages as they grow. Humans, like all other mammals, are born alive. When mammals are born, they can look like small, undeveloped versions of their parents. Many mammals need to be cared for by their parents until they can mature. Birds, fish, and most reptiles begin their life cycle by hatching out of eggs. When they hatch they look very much like their parents, except for size and color. Some animals, however, go through great changes during their life cycles.

Metamorphosis

Metamorphosis (met′ ə mor′ fə sis) is when animals change from one form to a completely different form during their life cycles. Butterflies, mosquitoes, wasps, fireflies, ladybugs, and bees go through a complete metamorphosis. There are four distinct stages in this metamorphosis.

1. Egg The fertilized egg is laid by the adult female.

2. Larva The egg hatches into a **larva**. It is a young animal that looks completely different from its parents. The larva eats and grows larger.

3. Pupa The larva stops feeding and enters into the **pupa** stage. During this stage, it covers itself with a special case. Many changes take place in the pupa.

4. Adult After going through changes, the adult comes out of the pupa. The adult can reproduce to start another life cycle.

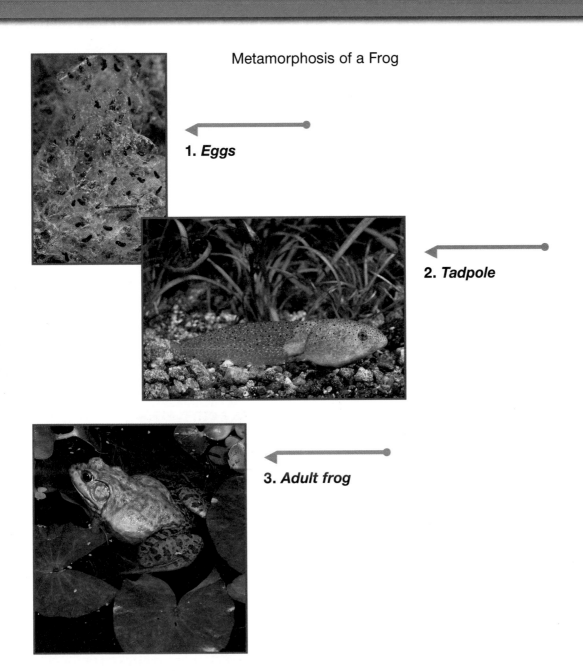

Metamorphosis of a Frog

1. *Eggs*
2. *Tadpole*
3. *Adult frog*

Some amphibians also go through a metamorphosis. A frog has a three-stage metamorphosis. First, a frog egg hatches into a swimming tadpole with gills and a tail. The tadpole grows front and back legs and the tail disappears. Once fully developed, it moves onto land, and the adult frog breathes with lungs instead of gills. The digestive system of the frog also changes so it turns from a herbivore to a carnivore during its metamorphosis.

Why We Care About Life Cycles

Knowing about life cycles can be helpful in controlling pests. Insects, for example, can be useful, but harmful insects can cause problems for farmers by damaging crops. Farmers can use biological methods, or natural ways, of killing harmful insects. In this way, they can avoid using chemicals that may be harmful to other organisms. One biological method is to interrupt the insect's ability to reproduce. To do that, scientists study the insect's life cycle.

There are also many other ways to use information about plant and animal life cycles. Have you ever eaten sweet corn, tomatoes, or bananas picked at just the right time in their life cycles? Nothing tastes better. Food producers know the right time to harvest during the plant's life cycle. If it is picked too soon, the food may not have its full nutritional value. If it is picked too late, the food may not be good to eat. But knowing the best time for harvesting crops is not the only reason for studying the life cycles. Scientists also study life cycles to learn how to live with other organisms that exist in the world we all share.

Knowing about different animal life cycles can help veterinarians and others who care for animals.

Knowledge about animal life cycles helps veterinarians and others who care for animals to provide for the animal's needs to survive. However, some needs can only be met by the young animals' parents. Many animals, such as puppies and kittens, depend on their parents for food during the first few weeks of life. The parents keep them clean and protect them. Kittens depend on their parents to teach them to catch mice. A lion teaches her cub to hunt. Young seals also learn to hunt from their parents.

If the young animals are separated from their parents too early, they will not learn these behaviors. This could mean that they may not be able to survive on their own. By studying life cycles, people with special training can sometimes help an orphaned animal to survive. This is especially important when the species is endangered.

A newborn kangaroo spends its first weeks of life in the pouch of its mother.

CHECKPOINT

1. What are the stages in a plant's life cycle?
2. How can animals change during their life cycles?
3. What are two reasons why we should care about plant and animal life cycles?

 How do plants and animals change as they grow?

ACTIVITY

Comparing Animal Life Cycles

Find Out
Do this activity to compare how long it takes for different animals to mature.

Process Skills
Communicating
Using Numbers
Inferring

WHAT YOU NEED

markers

posterboard

pens

pencils

reference books

Internet

glue

Activity Journal

WHAT TO DO

1. Work in groups and pick an animal to research. Do not pick the same animal that another group chooses.

2. Use references or the Internet to learn about your animal's life cycle. What are the different stages? How do the adults and young interact? How long does the animal live? How big does the animal get? Collect as much information about your animal as you can.

3. **Make a poster** showing the stages in your animal's life cycle. Use drawings and words to show its stages of development. On the left side of your poster, draw a picture of your animal at birth. On the right side of your poster, draw what your animal looks like as an adult.

4. With your class, make a bar graph titled "Average Animal Life Spans." Add a bar to the graph that shows how long your animal's average life span is.

5. Compare the average life spans of all the animals chosen by the class.

6. Share the information on your poster with the rest of your class and answer any questions they may have.

Conclusions

1. How is your animal different from the other animals chosen by your classmates?
2. Compare your animal's average life span with the others. Is it shorter or longer than the others?
3. How does your animal's size compare with the size of the other animals?

Asking New Questions

1. Do you think the size of an animal and its life span are related? Explain.
2. What are some animals that have a life cycle similar to your group's animal?

SCIENTIFIC METHODS SELF CHECK

✔ Did my group make a poster and share what we learned with the rest of the class?

✔ Did we **record** our animal's average life span on the class bar graph?

✔ Did we compare average life spans of animals?

A39

Carbon and Water Cycles

Find Out

- How plants and animals use carbon dioxide and oxygen
- What role plants play in the water cycle
- How humans influence Earth's cycles

Vocabulary

carbon dioxide
oxygen
carbon dioxide-oxygen cycle
water cycle
transpiration

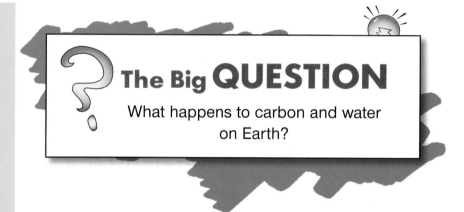

The Big QUESTION

What happens to carbon and water on Earth?

Carbon and water can both be found everywhere on Earth—in frozen glaciers, in dry deserts, and even at the bottom of the ocean. Why do you suppose this is true?

Carbon Dioxide and Oxygen

Just as plant and animal species continue to exist as long as their life cycles can continue, life on Earth continues because substances necessary for life are recycled over and over again. If these substances were not recycled, Earth's resources would have been used up millions of years ago. **Carbon dioxide** and **oxygen** are two gases that are necessary for organisms to live.

These gases are passed from one organism to another in both land and water environments. Think about a fish and some plants in a pond. Each uses some of what the other puts in the water. The plants use carbon dioxide from the fish, energy from the sun, and water during photosynthesis to make food. They also produce oxygen. The fish gets energy from eating plants or other animals. It uses oxygen that was produced by the plants and dissolved in the water. The plants also use some oxygen and produce some carbon dioxide as they use food. These exchanges between the fish and the plants happen over and over. The continuous exchange of oxygen and carbon dioxide makes up the **carbon dioxide-oxygen cycle.**

If you have ever had a fish tank, you have probably tried to balance the exchange of carbon dioxide and oxygen. A fish in a sealed container all by itself will eventually die as it uses up all of the oxygen in the water. But if the fish has the right amount of plants and food in its tank, it can live through its normal life cycle.

This fish will eventually use up all of the oxygen in this water.

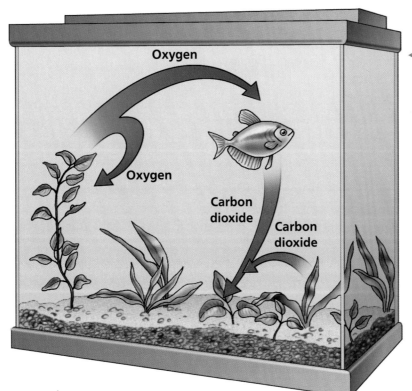

With the right exchange of carbon dioxide and oxygen, this fish will live through its normal life cycle.

The Water Cycle and Living Things

Life on Earth also depends on the cycling of water. Living organisms need water to survive, but they also help to recycle water. The circular pathway in which water moves through an ecosystem is the **water cycle.**

A Water Cycle

All living things need water. Green plants use water to make food. Some animals drink water. Other animals eat plants in order to get water. Still other animals live in the water.

When water falls to Earth, some of it runs along the ground and into oceans, lakes, rivers, and streams. Some water sinks into the ground. Later, it flows into lakes, rivers, and oceans.

Plants give off water from their leaves. This process is called **transpiration.**

When water vapor in the air cools, it changes back into liquid. It returns to Earth as dew, frost, rain, snow, sleet, or hail.

Water evaporates from the surfaces of lakes, streams, and oceans. As water evaporates, it becomes water vapor, a gas.

Humans and Earth's Cycles

The water cycle and the carbon dioxide-oxygen cycle are necessary for all forms of life on Earth. Human activities can greatly influence these cycles. For example, some carbon is locked up for a while in fuels. Coal and oil are fuels that were formed millions of years ago from the bodies of dead organisms. These fuels contain carbon. When humans burn these fuels, the carbon is given off to the air as carbon dioxide. Many people are concerned that we are releasing too much carbon dioxide into the air from burning these fuels.

Another way that humans can influence the cycles on Earth is by cutting down Earth's forests. This can result in less oxygen being released to the air during photosynthesis by the trees and other plants. Cutting down forests can also influence the water cycle in a variety of ways.

When you exhale, you are releasing carbon dioxide into the air.

Maintaining Balance

An important aspect to all cycles is that one part does not take over or become larger than the other part. Do you remember the fish in the bag of water at the beginning of this lesson? Without a plant in the bag to use up the carbon dioxide and to produce oxygen, the fish will die. The same thing can happen on Earth if cycles are interrupted. If more carbon dioxide is produced, the cycle may be changed. An interruption in a cycle can have many other effects. Understanding the balance in cycles can help us make wise choices for the future.

Cutting down forest areas can influence both the carbon dioxide-oxygen cycle and the water cycle.

CHECKPOINT

1. How do plants and animals use carbon dioxide and oxygen?
2. What role do plants play in the water cycle?
3. How can humans affect Earth's cycles?

 What happens to carbon and water on Earth?

ACTIVITY

Observing Transpiration

Find Out
Do this activity to learn how plants recycle water.

Process Skills
Predicting
Observing
Communicating
Inferring
Designing Investigations

What You Need

small potted plant

tape or string

water

clear plastic bag

Activity Journal

What to Do

1. Water the plant well. Place the plastic bag over the green leaves of the plant and gently close the bag around the stem. Be careful not to break the leaves or the stem.
2. Tie the plastic bag shut at the stem with the tape or string.
3. Place the plant in a sunny location for several hours.

4. **Predict** what you think will happen to the plant leaves. **Write** your prediction.
5. After several hours look at the bag and the leaves. **Observe** what is inside of the bag. **Record** your observations.

CONCLUSIONS

1. What happened in the bag?
2. Why did this happen?

ASKING NEW QUESTIONS

1. **Predict** what would happen to the plant leaves if you kept the plastic bag on the stem.
2. **Predict** what would happen if you moved the plant into a dark area.
3. **Design investigations** to test your predictions.

SCIENTIFIC METHODS SELF CHECK
✔ Did I **predict** what would happen to the plants?

✔ Did I **test** my predictions?

✔ Did I **observe** and **record** what happened?

✔ Did I **infer** why there was moisture in the bag?

Lesson 3

Energy Flow

Find Out
- What food chains are
- How energy from the sun is transferred to plants and animals
- What happens to energy when it moves through a food web

Vocabulary
food web
producers
consumers
decomposers
energy pyramid

The Big QUESTION

How does the sun influence the food we eat?

*D*id you drink some milk today? Or eat a salad or a chicken sandwich? Foods like milk, salad, and chicken sandwiches give you energy to work and play. But did you know that the sun is also responsible for the energy you need to play?

Food Chains

When plants produce sugars during photosynthesis, they are taking energy from one source and changing it to a type of energy they can use. When animals eat these plants, they are also taking energy from the plants and changing it to energy they can use. The energy from the sun flows to plants and then to animals. The main source of energy on Earth that is used by living organisms is the sun.

How does energy flow? One living organism may become food for another living organism. The energy in food is transferred from one organism to the next in the food chain. Look at the diagram. The arrows show how energy moves through the system.

Overlapping food chains are called a **food web.** Any organism in a food chain can eat or be eaten by many different organisms. Can you think of some different food chains that might exist in the food web shown here?

You are also part of the flow of energy on Earth. Think about what you had for lunch yesterday. If you had a salad for lunch, the energy you received came from the lettuce, tomatoes, and other things in your salad. Those plants received their energy from the sun during photosynthesis. But what if you didn't eat any plants for lunch yesterday? You still played a part in the flow of energy on Earth. If you had a glass of milk and a hamburger, the energy you received came from a cow and a steer. They received their energy from the grass or grains that they ate. And those grasses or grains received their energy from the sun. The energy is changed by each organism as it passes through the system.

A Food Web

Energy Moves Through the System

Producers

Plants and animals play different roles in the transfer of energy on Earth. Green plants are called **producers** because they use the sun's energy, along with other substances, to make their own food. Energy from the food is used to keep the plants alive. Plants also store some of the energy in the form of sugars in their roots, stems, and leaves.

Consumers

Many organisms can't make their own food. They are called **consumers** because they get their energy from plants or other consumers. Examples of consumers are rabbits, snakes, wolves, seagulls, and whales. Consumers rely on producers for the energy they need. Consumers need the energy to maintain life. They also store energy in their muscles and in other body tissues.

Some consumers are called scavengers. You may have seen scavengers eating animals killed on highways. Vultures, crabs, catfish, and hyenas get their energy by eating dead organisms.

Wheat plants and orange trees are examples of producers.

Rabbits and snakes are consumers.

Decomposers

Some consumers are called decomposers (dē′ kəm pōz′ ərz). **Decomposers** get energy by consuming dead organisms. Examples of decomposers are fungi, such as molds and mushrooms. Some worms are also decomposers. However, many decomposers, like some forms of bacteria, are so small that you cannot see them without a microscope. However, bacteria are the most abundant forms of life on Earth. Even though they can be very tiny, decomposers are important. When plants or animals die, their bodies are recycled and water, carbon, and other substances are not lost. Decomposers have the ability to break the body matter down.

Scavengers, like these vultures, are consumers too.

Have you ever left some strawberries in the refrigerator too long? If you have, you may have seen mold at work consuming the fruit. The mold can break down the strawberries into simpler substances. In this way, the substances can be returned to the soil and the air. Decomposers like mushrooms that grow on rotting logs in the forest help to break down the log so the carbon and all of the other materials can be recycled. These substances will be used by new plants. Without decomposers, the matter in living organisms would not be able to be recycled.

Mushrooms and molds are decomposers.

Energy Available

How much energy does an organism gain when it consumes another? A pyramid is a good model to show the flow of energy because a pyramid gets smaller as it goes up. The **energy pyramid** shows the amount of energy flowing from one organism to the next through a food chain. How many organisms are shown at the top of this energy pyramid? Only one consumer—the wolf—is at the top of the pyramid because the energy available to different organisms decreases every time it is transferred.

When producers get energy from the sun, some of it is used and only part of it is passed along to a consumer. When consumers eat the plants, again some of the energy is used and some of it is passed along to the next consumer. This continues with every level of consumer.

Because less of the energy is available from consumers higher on the pyramid, more of them need to be eaten. The animals on the top of the energy pyramid need lots of producers and herbivores to stay alive. That is one reason why there are more rabbits in a forest than foxes. That is also why there are more sardines in an ocean than there are sharks. A change in the number of producers will affect all the organisms in the food chains and webs.

The Energy Pyramid

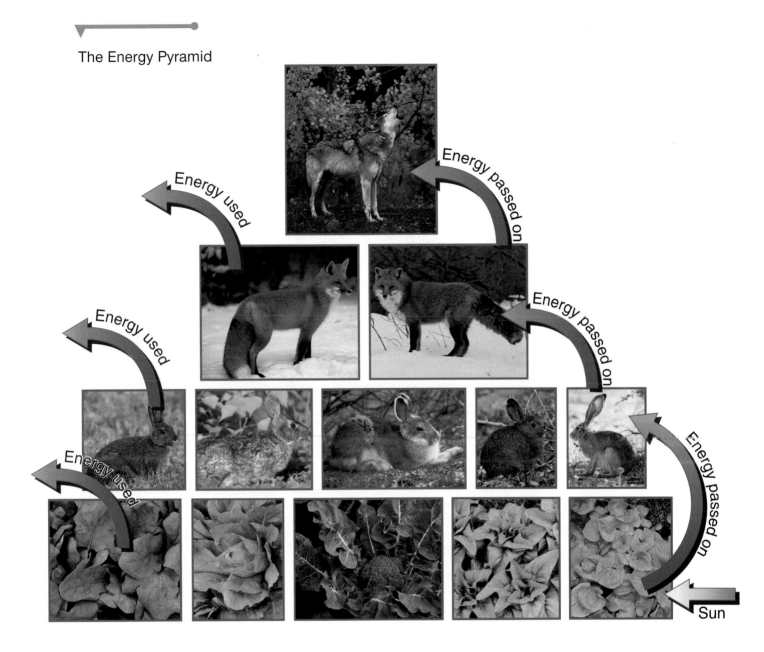

CHECKPOINT

1. What are food chains and food webs?
2. How is energy from the sun transferred to plants and animals?
3. What happens to energy as it moves through a food web?

 How does the sun influence the food we eat?

ACTIVITY

Food Webs

Find Out
Do this activity to see how a change in one link of a food chain can affect all the other links.

Process Skills
Constructing Models
Inferring

WHAT YOU NEED

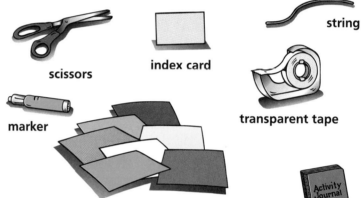

scissors, index card, string, marker, transparent tape, colored construction paper (green, white, red, orange, yellow, blue, and brown), Activity Journal

WHAT TO DO

1. Set the orange sheet of paper aside. Trace the index card on the other sheets, making at least six of each color (the more the better).

2. Write these words on the cards:
    on green—*plants;*
    on white—*herbivores;*
    on red—*predators;*
    on blue—*omnivores;*
    on yellow—*scavengers;*
    on brown—*decomposers.*

3. Cut a large circle out of the orange paper and label it "Sunlight."

4. Arrange the green plant cards around the orange circle. Tape a small piece of string between each card and the circle. Tape herbivore cards to plant cards. Then, tape predator cards and/or scavenger cards to herbivore cards. Attach decomposer cards to any of the other cards. When you are finished, you will have made a **model** of a food web.

5. Now, remove half of the green cards. Take the other cards that were attached to them and try to connect them somewhere in your web. All of the animal cards need a food source, but you can't attach more than two cards to the same food source.

CONCLUSIONS

1. What happened to the food web when some of the cards were removed?
2. What would happen in nature if one of the animal cards were not attached to a food card?

ASKING NEW QUESTIONS

1. What organisms do you depend on in your food web?
2. What would happen to your web if the orange sun circle were not there?

SCIENTIFIC METHODS SELF CHECK
✔ Did I **make a model** of a food web?
✔ Did I **infer** what would happen to consumers if producers were destroyed?

Review

Reviewing Vocabulary and Concepts

Write the letter of the answer that completes each sentence.

1. All the stages in an organism's life are called its ___.
   - **a.** life cycle
   - **b.** carbon cycle
   - **c.** water cycle
   - **d.** pupa

2. We call the sprouting of seeds ___.
   - **a.** germination
   - **b.** oxygen cycle
   - **c.** carbon dioxide
   - **d.** nutrients

3. After emerging from an egg, a frog is in the form of ___.
   - **a.** a larva
   - **b.** oxygen
   - **c.** a tadpole
   - **d.** a pupa

4. The stage that follows the larva in the life cycle of a butterfly is ___.
   - **a.** carbon
   - **b.** pupa
   - **c.** web
   - **d.** chain

5. The amount of energy flowing from one organism to the next through a food chain is called the ___.
   - **a.** ecosystem
   - **b.** life cycle
   - **c.** producers
   - **d.** energy pyramid

6. The movement of carbon dioxide from animals to plants is part of the ___.
   - **a.** energy flow
   - **b.** germination
   - **c.** carbon dioxide-oxygen cycle
   - **d.** metamorphosis

7. We call overlapping food chains ___.
   - **a.** a balanced diet
   - **b.** a food web
   - **c.** a carbon cycle
   - **d.** energy

8. Organisms that make their own food are called ___.
   - **a.** consumers
   - **b.** producers
   - **c.** transpiration
   - **d.** energy

9. Molds and mushrooms are examples of ___.
   - **a.** producers
   - **b.** omnivores
   - **c.** consumers
   - **d.** decomposers

Understanding What You Learned

1. List four stages of plant growth and what happens during each stage.
2. How do food growers decide when to harvest crops?
3. How do organisms that were alive millions of years ago take part in the carbon dioxide-oxygen cycle?

Applying What You Learned

1. Explain what metamorphosis is and give one example of it.
2. How is the way decomposers get energy different from the way producers get energy?

3. What do all living organisms need? Explain why this is necessary for life.

For Your Portfolio

Draw an energy pyramid that includes you. Think about the foods you have eaten in the past few days. Try to use some of those foods in your energy pyramid. Where would you be in the energy pyramid?

Review

A57

CHAPTER 3

Ecosystems

Freshwater and oceans cover most of Earth's surface, and many organisms live there. These areas are known as aquatic ecosystems. Perhaps there's a body of freshwater near you. A river might be cold and fast-moving at its source. Moss and algae might grow on rocks in the water. Perhaps you can see trout or other fish.

Downstream you might find fish such as carp or catfish. Cattails and willow trees may grow along the banks. Where the river meets the ocean, you can see mudflats or salt marshes. Where the ocean tides wash over the shore, tidal pools can form. Many different living organisms can be found in all of these aquatic ecosystems.

The Big IDEA

An ecosystem represents the interactions between living and nonliving things.

CHAPTER SCIENCE INVESTIGATION

Nonliving factors can influence the organisms living in an ecosystem. Find out how in your *Activity Journal.*

LESSON 1

Characteristics of Ecosystems

Find Out
- How big an ecosystem is
- How resources affect an ecosystem
- Why animals leave an ecosystem

Vocabulary
ecosystem
competition
niche
migration
barriers

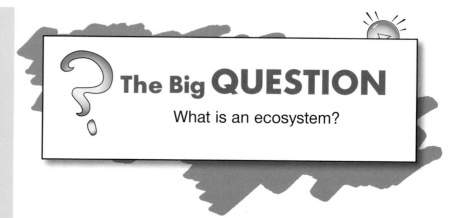

The Big QUESTION

What is an ecosystem?

Picture a mountain meadow; an elk grazes among meadow flowers. You decide to study the mountainside, its meadows, flowers, elk, and all living and nonliving things on it. What would you be studying?

Ecosystems

Living organisms in a community depend on one another. The plants and animals act upon—or interact with—each other. For example, some animals are herbivores that eat the plants in an area. Other animals are carnivores that eat the herbivores or other carnivores. When any of the animals die, scavengers come and eat the remains of the dead animal. Decomposers also break down the remains, enriching the soil so that new plants can grow.

An ecosystem can be small.

All of the living organisms in a community depend on the actions of the other organisms. However, the living things in a community also interact with many nonliving things in their surroundings such as sunlight, water, soil, and temperature. How much water is available can influence what types of plants grow in an area. The air temperature can determine if a species of animal will survive in the area. In order to understand why plants grow in a certain area, we have to look at many factors. When we study all of the interactions between the living and the nonliving things in an environment, we call that environment an **ecosystem.**

The term *ecosystem* was first used in 1935 by an English scientist named Sir Arthur Tansley. When a person decides to study an area, he or she decides how large or small the ecosystem will be. There are ecosystems all around us. Ecosystems can be as large as an ocean or as small as a puddle. Whatever the size, when the nonliving factors change, the groups of living organisms in an ecosystem also change.

An ecosystem can be large. No matter what the size, an ecosystem is made up of all living and nonliving things in the area.

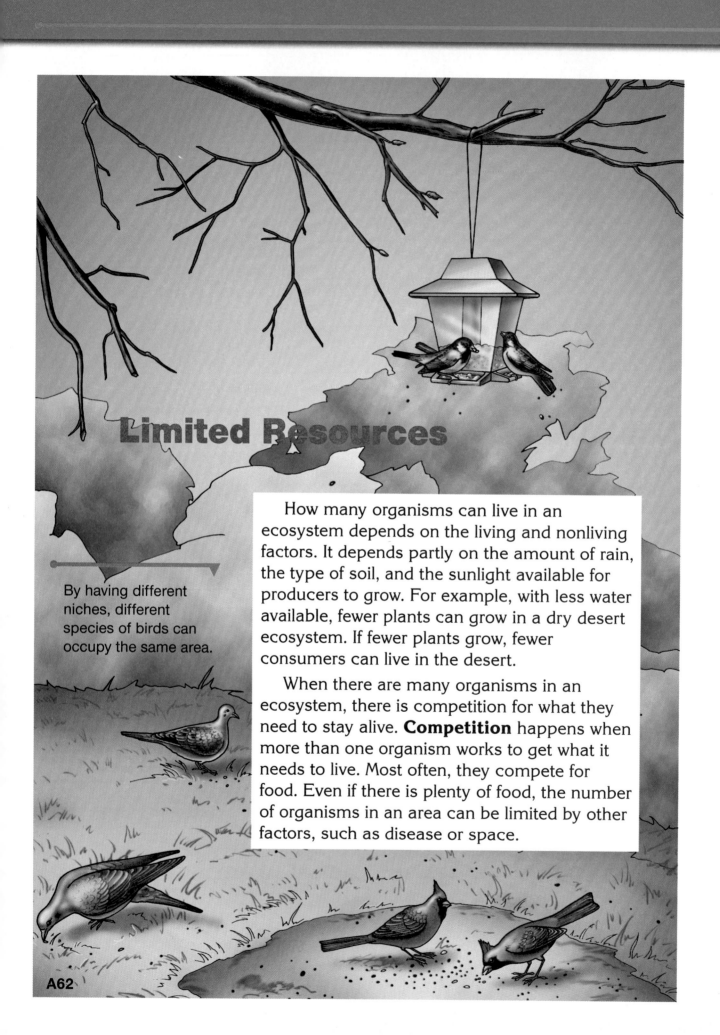

Limited Resources

By having different niches, different species of birds can occupy the same area.

How many organisms can live in an ecosystem depends on the living and nonliving factors. It depends partly on the amount of rain, the type of soil, and the sunlight available for producers to grow. For example, with less water available, fewer plants can grow in a dry desert ecosystem. If fewer plants grow, fewer consumers can live in the desert.

When there are many organisms in an ecosystem, there is competition for what they need to stay alive. **Competition** happens when more than one organism works to get what it needs to live. Most often, they compete for food. Even if there is plenty of food, the number of organisms in an area can be limited by other factors, such as disease or space.

Competition does take place but living organisms avoid it. When that is possible, it helps the organism to survive. Having a particular behavior or food gives organisms their own place in the ecosystem. This place is their niche. A **niche** (nich) is the organism's role in the ecosystem. By having different roles, many different plant and animal species can occupy the same area.

For example, if you look at a bird feeder in a park you will see many different birds there. However, because of their different niches, these birds may not be competing. Sparrows, doves, and cardinals all eat different kinds of seed. The small sparrows eat the tiny thistle seeds from the holes in the feeder. As the sparrows kick out the bigger seeds, the cardinals collect the hard-shelled sunflower seeds from the ground. And the doves pick up any remaining seeds from the area. The different niches of the birds increase their chances for survival.

Each species of plant also has its own role, or niche, which allows it to live and grow. For example, the huckleberry bush and the spruce tree can both grow in rocky soil that is very acidic. Many other species of plants cannot grow in this type of soil. Because of their soil requirements, both the huckleberry and the spruce appear to compete for the same nutrients. However, they have slightly different needs for sunlight. The huckleberry plant is only ½ m high, but it can grow in some shade. Spruce trees grow much taller and shade the huckleberry. The different niches of the huckleberry and the spruce tree allow them to survive in the ecosystem.

Hawks and owls have the same prey, but hawks hunt during the day and owls hunt at night, so they have different niches.

Huckleberry plants and spruce trees both grow in rocky soil but they have different sunlight needs.

Migration

Increasing the numbers of a species in an ecosystem can be an adaptation for survival. Clumping happens when organisms form groups for protection or because resources such as water or food are mostly found in one spot. Wolves hunt together in a pack. Cottonwood trees grow together along streams because a resource they need—water—can be found there. Penguins huddle together by the thousands to conserve body heat in very cold climates. By increasing their numbers, the living organisms improve their use of the nonliving factors in their environment.

However, sometimes the number of organisms can grow too large. When too many organisms compete for the same resources, diseases and hunger may spread, causing some to die. When this happens, some animals leave the area, increasing their chances for survival. **Migration** is moving to a new area that often has more room and more resources.

Some species have the instinct to migrate at certain times in order to reproduce or because of weather conditions or sunlight changes. Other animals may migrate when food sources are low. Every three or four years, lemming populations become so large that they migrate in search of food. Wildebeests and zebras migrate after heavy rains. By migrating with the change of seasons, these species increase their food supplies and decrease their competition for food and space.

Canadian geese migrate with light and weather changes when food sources may become low.

Barriers to Migration

Sometimes organisms cannot migrate because their path is blocked. **Barriers** to migration stop movement from one area to another. Natural barriers might be bodies of water, such as lakes, rivers, and oceans. Other barriers could be mountains or deserts. Extreme hot or cold temperatures can stop some species from migrating, as can heavy snowfall or harsh winds. When organisms cannot migrate, they may face death.

Humans can also create barriers to migration. Dams on rivers can stop animals from migrating up or down streams. Highways can block migration paths or create dangerous obstacles for animals to cross. New housing and construction can isolate plants and animals so they cannot move from one area to another. The human impact on an area can change the makeup of the ecosystem. Sometimes, what humans do can endanger an entire species of plant or animal. We need to consider how our actions can impact the living organisms that share our ecosystems.

Roads and housing developments can create barriers to migration for some species.

CHECKPOINT

1. How big is an ecosystem?
2. How can resources affect an ecosystem?
3. Why do some animals leave an ecosystem?
4. What is an ecosystem?

ACTIVITY

Influencing an Ecosystem

Find Out
Do this activity to learn how nonliving things can change an ecosystem.

Process Skills
Measuring
Controlling Variables
Hypothesizing
Communicating
Observing
Inferring

What You Need

two 260-mL cups

marking pen

two plant cuttings

small spoon

graduated cylinder

Activity Journal

salt

What to Do

1. **Mark** one cup with an "S" and **put** one rounded spoonful of salt into it. Fill with 100 mL of water and stir until the salt dissolves.

2. Fill the second cup with 100 mL of water.

3. Put a plant into each cup. Write a hypothesis about what will happen to the two plants.

4. Place both plants in the same location. After three days, observe the plants. Write down your observations.

Conclusions

1. Compare your hypothesis with your observations.
2. What can you infer about the plant placed in salt water?
3. Why did the two plants react differently?

Asking New Questions

1. Predict how salt might affect other plants. Test your prediction.
2. Can farmers use ocean water to water their crops? Why or why not?

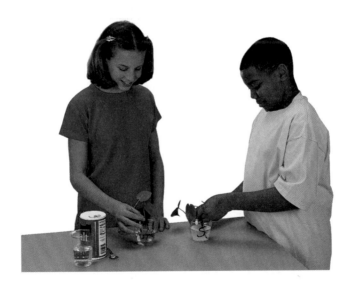

SCIENTIFIC METHODS SELF CHECK

✔ Did I **put** the plants in the same amount of water?

✔ Did I add salt to the water for just one plant?

✔ Did I test my **hypothesis**?

✔ Did I **infer** what had an influence on that plant?

Lesson 2

Ocean Ecosystems

Find Out
- How ocean fish are different from freshwater fish
- How minerals are recycled in the ocean
- What lives in the ocean's deep layers

Vocabulary
phytoplankton
zooplankton
sunlit zone
twilight zone
dark zone
abyss

The Big QUESTION

What makes up ocean ecosystems?

Remember that a person studying an ecosystem determines its boundaries. Since 75 percent of Earth is covered by water, many thriving ecosystems can be found under water. Even the deepest parts of the ocean have living organisms interacting with their environments.

Ocean Animals

Earth's oceans support life in a wide variety of ways. Many different ecosystems can be found in the different layers of the oceans, but all of these systems interact in a balanced way to support the cycles of life in the water. Let's examine the organisms in an ocean to see how these interactions take place.

Ocean animals have characteristics that allow them to survive. They have adapted to the high levels of salt in their environment. For example, because saltwater fish lose body fluids through their skin and gills, they drink a lot of water. They must drink water. However, freshwater fish are different. They absorb, or take in, water through their skin and gills. Freshwater fish don't need to drink water.

The streamlined shapes of many fish in the ocean help them to move quickly and easily through the water. But some ocean animals that do not have the ability to swim can float with the currents. The currents carry them from place to place. Other animals attach themselves to the floor of the ocean or rocks and do not move. These animals depend on currents to carry food to them. Animals in the ocean also protect themselves from predators in a variety of ways.

The purple sea urchin is covered with long, movable spines. These sharp spines keep other animals from eating the urchin.

Jellyfish float with the currents. Tiny zooplankton drift along with the jellyfish.

The lionfish has long, colorful fins that contain deadly poison. They use these fins to attack other fish.

An Ocean Food Web

Many of the ocean's producers live in the sunlit surface layers of water. The water here is warmer than it is in lower layers because of the warm air above the water. Plankton are tiny plantlike organisms called **phytoplankton** and animal-like organisms called **zooplankton** that float at or near the water's surface. During photosynthesis, phytoplankton and other plantlike organisms use sunlight, carbon dioxide, water, and nutrients to produce food. The zooplankton eat the tiny phytoplankton. These plankton become food for fish and other marine animals.

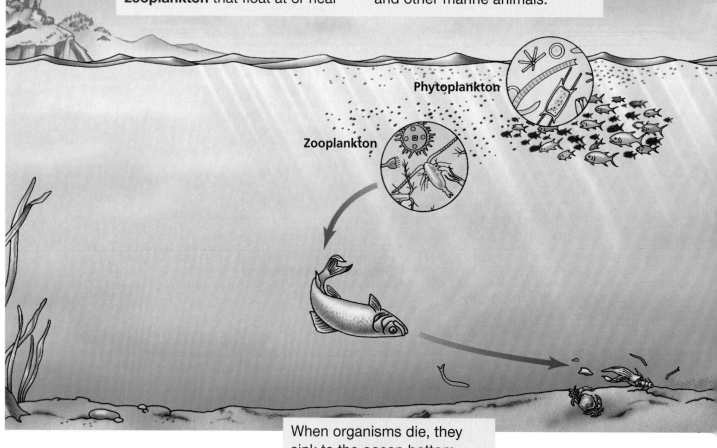

When organisms die, they sink to the ocean bottom. They are usually eaten by scavengers and decomposers who live in the ocean's lower depths. The dead organisms that are not eaten decay. During this process, mineral salts are returned to the ocean.

Since the water in the ocean is constantly moving, the rising currents carry the minerals back to the sunlit surface, where producers will again use them to make food and to release oxygen into the water.

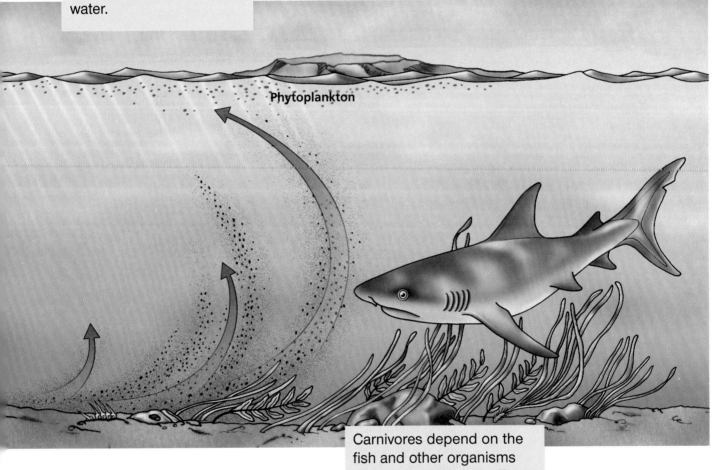

Phytoplankton

Carnivores depend on the fish and other organisms who eat the plankton. There are fewer sharks than other fish because sharks are on the top of the energy pyramid in the ocean's ecosystem.

The Ocean's Layers

Remember that ecosystems are the interactions of living and nonliving things that we study in an environment. Scientists are beginning to understand how the nonliving parts of the ocean change as you travel deeper. They have divided ocean layers into zones. The upper layers are in the **sunlit zone,** where sunlight shines and seasonal changes influence the water temperature.

Below the sunlit zone is the **twilight zone,** where the water is much colder and the water pressure is much greater. There is so little light in the twilight zone that plants cannot carry out photosynthesis. Beyond the twilight zone is the **dark zone,** where there is no light. The water is very, very cold and the water pressure severe. Some bacteria can be found in this area, but animals that live here must depend on sounds to locate one another because no light can reach this area.

The deepest areas of the ocean are called the **abyss** (ə bis′). Scientists are still working to find out more about interactions of living and nonliving things in the abyss.

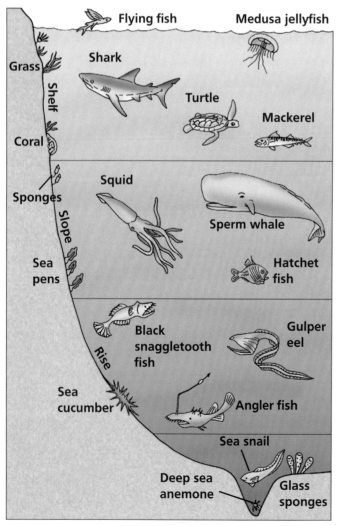

Different Layers in the Ocean

Until recently, we knew very little about organisms living in the lower depths of the ocean. Organisms of the deep ocean live under conditions different from the living conditions of other organisms. People once thought that nothing could survive in the water of the deep ocean. No one believed that organisms could withstand the extreme water pressure and the lack of light. Scientists now know that bacteria and other organisms have adapted to life in the lower depths.

The pineapple fish stays on the ocean bottom during the day and travels toward the surface at night.

The deep sea hatchetfish has rows of light organs along the lower edge of each side of its body.

The gulper eel can open its large jaws to catch prey larger than itself.

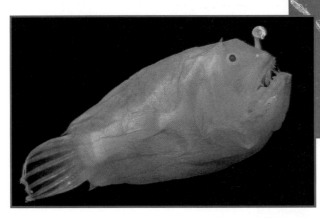

The tip light of the deep sea angler fish attracts its next meal.

CHECKPOINT

1. How are ocean fish different from freshwater fish?
2. How are minerals recycled in the ocean?
3. What lives deep in the ocean?
 What makes up ocean ecosystems?

ACTIVITY

Decomposers

Find Out
Do this activity to see how nonliving factors influence decomposers.

Process Skills
Controlling Variables
Predicting
Communicating
Observing
Inferring
Designing Investigations

What You Need

two bananas

one plastic knife

two packets of dry yeast

four resealable sandwich bags

tap water

a marker

Activity Journal

What to Do

1. Use the marker to label the bags "1," "2," "3," and "4."
2. Slice the bananas into equal pieces. Put three slices of banana into Bag 1.
3. Put three slices of banana and empty a packet of yeast into Bag 2.
4. Put three slices of banana and some water into Bag 3.

5. Put three slices of banana, some water, and empty a packet of yeast into Bag 4.

6. Put the bags in a sunny place and leave them for two days.

7. After two days, **observe** what happened in each bag. Compare the bags from different groups. **Record** your observations.

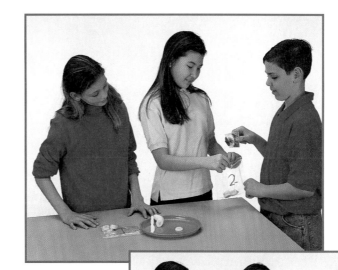

CONCLUSIONS

1. What happened to the bananas in each bag?
2. Why are the bananas in each bag reacting differently?

ASKING NEW QUESTIONS

1. **Predict** how the bananas and yeast would react in the dark or in the cold. **Design an investigation** to test your prediction.
2. Why do you think bakers use yeast when baking bread?

SCIENTIFIC METHODS SELF CHECK

✔ Did I **observe** and **record** what happened in each bag?

✔ Did I put the same number of banana slices into each bag and seal the bag?

✔ Did I **infer** what had happened in Bag 4?

A75

Lesson 3

Changing Ecosystems

Find Out
- How ecosystems change
- Why plants grow well in the pioneer stage
- How humans can change an ecosystem

Vocabulary
succession
pioneer stage

The Big QUESTION

Why do ecosystems change?

Do you know of a vacant lot, a patch of ground that has not been cultivated, or even cracks in a cement or paved surface near you? If you look closely in any of these places, you may find niches for living things. But niches do not always stay the same. Ecosystems all around you are constantly changing.

An Ecosystem Changes

All ecosystems tend to change, although the speed of change can vary greatly. Fire can cause the parts of an ecosystem to change quickly; organisms must then adapt, migrate, or die. However, changes may also be slow and sometimes go unnoticed. Changes may be happening in the ecosystems near you.

Have you ever seen an abandoned building or lot? When people abandon a building, other organisms can begin to live there. Wild grasses, shrubs, and weeds can begin to germinate in undisturbed soil. Fungi may cause the wooden boards to rot. Earthworms may live in the soil and decompose the pieces of wood. Bacteria can grow and break down the materials further. Even weathering and erosion can help to break down the rocks.

With producers growing in the area, consumers will come to feed. Small mice and rodents can feed on the weeds and grasses. Insects will eat the plants and other insects. Birds may also come to feed on the insects and seeds. As more animals and plants die and decompose, more nutrients are added to the soil, creating a rich environment for more producers to grow.

Changes in an ecosystem can be very slow.

The Stages of Change

New plant and animal populations replace old plant and animal populations over time. This replacement is called **succession.** Why does succession happen? Nonliving factors such as the climate, soil, and nutrients available in an area determine the types of organisms that can survive there. If one of the factors changes, the whole ecosystem will change.

If a forest area suddenly catches fire, many of the plants will not be able to live there. If the plants die, the animals who depend on those plants for food and shelter will also no longer be able to live in that ecosystem. New species that can survive in a burned-out area will replace the old plant and animal species.

Succession is a normal stage in the development of many ecosystems. Without succession, living organisms would not be able to move into new areas. When farmers clear grassland and plant it with crops, the organisms living in the area will change. However, what will happen if the farmers move out of the area and abandon their farms? Again, the organisms in the area will change; the crops will no longer grow as well. Eventually, they will be replaced by tall grasses and other plants, mice, rabbits, insects, and birds. This first stage of change is called the **pioneer stage.** During the pioneer stage, the plants grow very well because there is little competition.

As more plants grow, the competition becomes stronger. The taller plants such as young trees begin to shade the shorter plants, blocking the sunlight they need to grow. As the competition for sunlight increases, the taller trees will replace the grasses. With trees come larger animals like deer and squirrels. Predators, such as hawks and owls, can live in the trees and hunt the mice for prey. As the trees grow, the ecosystem now looks very different from the farm.

Eventually the ecosystem will reach a balanced stage. Changes will occur only if the living or nonliving factors again change.

Succession is a normal stage in the development of an ecosystem.

It takes about 70 years for a balanced forest ecosystem to replace farmland.

Change Can Threaten Species

Change is a part of any ecosystem, but sometimes the actions of humans can bring about change very quickly. When their habitats change quickly, species of animals and plants can be threatened with extinction. When a species becomes extinct, no changes can bring the animals or plants back. Since each plant and animal is a part of the balance of living things, its extinction can threaten the survival of other living organisms, including people. The plants and animals here are endangered or threatened species.

In the mid-1850s about 20 million North American bison roamed the western plains. By the 1880s only 551 were left. Efforts were made to protect this animal, and today about 15,000 bison live in wildlife preserves.

Darwin's rheas live on the plains of South America. Habitat destruction has endangered this animal.

The North American pitcher plant grows in acidic soils. The pitcher plant traps insects inside of its long, tubelike leaves and slowly digests them. It is endangered because of habitat destruction.

The loggerhead turtle lays eggs on ocean shores. It has become endangered because of beach development, habitat destruction, and pollution.

The Madagascar ring-tailed lemur lives in the forests of Madagascar. Because of deforestation, all species of lemurs are endangered.

The wild Asian yak inhabits the cold, dry plateaus of Tibet. Because of overhunting, it has almost become extinct.

CHECKPOINT

1. How do ecosystems change?
2. Why do plants grow well in the pioneer stage?
3. How can humans change an ecosystem?
 Why do ecosystems change?

ACTIVITY
Adding to Ecosystems

Find Out
Do this activity to find out how earthworms can change the soil of an area.

Process Skills
Hypothesizing
Communicating
Observing
Inferring
Designing Investigations

What You Need

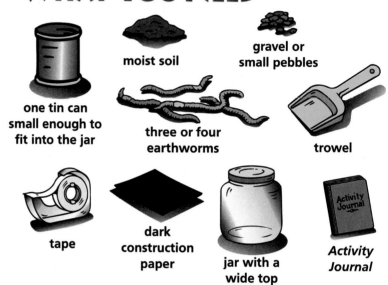

one tin can small enough to fit into the jar

moist soil

three or four earthworms

gravel or small pebbles

trowel

tape

dark construction paper

jar with a wide top

Activity Journal

What to Do

1. Put a layer of gravel in the jar. Then place the can into the jar.
2. Use the shovel to carefully place soil between the can and the glass up to the height of the can.
3. Gently place the worms on top of the soil. Tape the construction paper around the sides of the jar to keep the light out.
4. Write a hypothesis about what will happen to the worms.

Safety! *Wash your hands after handling soil and worms.*

5. After two days, take the paper off the jar. **Observe** what has happened and **write** down your observations. Make sure the soil stays moist. Add a little water if needed.

6. Cover up your jar for a couple of days and then observe it again. What else has happened? When you are finished, release the worms outside in an area that has soil.

CONCLUSIONS

1. Compare your **hypothesis** and your observations.
2. Where did the worms go in the jar? Why?
3. How are the worms changing the soil?

ASKING NEW QUESTIONS

1. Why are worms valuable for farmers?
2. How can earthworms help to change a vacant lot in a city?
3. Write another question you have about earthworm behavior. **Design an investigation** to find an answer to your question.

SCIENTIFIC METHODS SELF CHECK
✔ Did I test my **hypothesis**?
✔ Did I **observe** what happened in the jar?
✔ Did I **write** down what happened?
✔ Did I **infer** how earthworms can help to change an ecosystem?

Review

Reviewing Vocabulary and Concepts

Write the letter of the answer that completes each sentence.

1. All living and nonliving things that interact within an area under study are called ___.
   a. a competition b. a pioneer stage
   c. a migration   d. an ecosystem

2. An organism's particular role where it does not have to compete for food is its ___.
   a. abyss   b. succession
   c. twilight zone d. niche

3. Tiny organisms that make their own food and float near the ocean's surface are ___.
   a. phytoplankton b. pioneer stage
   c. succession   d. zooplankton

4. The ocean's upper layer is called the ___.
   a. competition   b. sunlit zone
   c. zooplankton   d. twilight zone

5. The replacement of plants and animals over time within an ecosystem is called ___.
   a. competition   b. dark zone
   c. a niche   d. succession

Match the definition on the left with the correct term.

6. when an organism works against another in its environment for space or food
   a. migration

7. the movement to a new place to find more space and resources
   b. extreme water pressure

8. a nonliving factor that organisms living deep within the ocean must adapt to
   c. competition

A84

Understanding What You Learned

1. Who decides the size of an ecosystem?
2. What are two reasons an animal migrates?
3. Name one adaptation organisms that live in the ocean have.

Applying What You Learned

1. Explain how some natural events could cause an ecosystem to change rapidly.
2. Can a fish from a pond live on the bottom of the ocean? Why or why not?
3. Explain how human beings can change an ecosystem.

4. What happens if you change something in an ecosystem?

For Your Portfolio

Choose an ecosystem in your area and describe all of the interactions that might take place. Make sure to tell what the boundaries are for your ecosystem. Include living and nonliving factors such as temperature, light, etc.

Unit Review

Concept Review

1. Explain how plant and animal structures make it possible for them to survive.

2. How do plants and animals get the energy they need to grow?

3. Think of an ecosystem such as a park or a vacant lot. Explain how the nonliving things make it possible for plants and animals to live there, and how the living organisms interact with each other.

Problem Solving

1. Select an animal and a plant that live in the same aquatic ecosystem. Explain what each would need to do to survive if there were a big change in the water temperature in the ecosystem.

2. Make a list of five plants and five animals that live in one ecosystem. Draw a food web that shows who eats what. Explain what would happen if one of the plants or plant-eating animal species suddenly became extinct.

3. Imagine that you are a park ranger in Africa. Your park is trying to build a road to increase tourism and to raise money for the park. However, the plans for the road will interrupt the migration path for an endangered species of lizard that lives only in the park. What can you do?

Something to Do

Select a small animal and plan a home for it that would provide for as many of its needs as possible, without human intervention. List the nonliving and living essentials the animals would need over time. Explain what might have to be provided by a human being and why that is so.

UNIT B

Earth Science

Chapter 1 **The Solar System** **B2**
    Lesson 1: **Gravity and the Sun** B4
        Activity: Modeling the Solar System B12
    Lesson 2: **The Moon and Earth** B14
        Activity: Gravitational Pull B20
    Lesson 3: **Space Exploration** B22
        Activity: Making a Balloon Rocket B28
    Chapter Review **B30**

Chapter 2 **Weather** **B32**
    Lesson 1: **Air and Wind** B34
        Activity: Making Magic with Air B40
    Lesson 2: **Water Cycle and Clouds** B42
        Activity: Making a Cloud B50
    Lesson 3: **Severe Weather** B52
        Activity: Making a Barometer B58
    Chapter Review **B60**

Chapter 3 **Shaping and Reshaping Earth** **B62**
    Lesson 1: **Weathering** B64
        Activity: Ice Power B70
    Lesson 2: **Erosion** B72
        Activity: Eroding with Water B78
    Lesson 3: **Catastrophic Events** B80
        Activity: Making an Earthquake B86
    Chapter Review **B88**

Chapter 4 **Rocks, Minerals, and Soils** **B90**
    Lesson 1: **Rock Types and Formation** B92
        Activity: Classifying Rocks B98
    Lesson 2: **Earth's Minerals** B100
        Activity: Naming Minerals B106
    Lesson 3: **Soils** B108
        Activity: Testing Soil Characteristics B114
    Lesson 4: **Soil as a Natural Resource** B116
        Activity: Modeling Earth's Limited
                 Resources B122
    Chapter Review **B124**

Unit Review **B126**

CHAPTER 1
The Solar System

Three, two, one, liftoff! Imagine you are an astronaut. You've just been launched into space. At first, your body is pushed back in your seat. It's as if the weight of the world is sitting on you. Then, when the rocket engine stops, you start to feel weightless. Things begin to float in the spacecraft. When you look out the spacecraft's tiny window, you see stars all around you. In the distance, you see Earth getting smaller and smaller.

Have you ever daydreamed about what it would be like to travel on a rocket ship? What would you do on your space trip? What would it be like to live without Earth's gravity? Where would you like to go?

The Big IDEA

Gravity is a force that exists between objects.

CHAPTER SCIENCE INVESTIGATION

Make a model of a moonscape. Find out how in your *Activity Journal*.

Lesson 1

Gravity and the Sun

Find Out
- What makes up the solar system
- What the different planets are
- How long it takes the planets to travel around the sun

Vocabulary
gravitational pull
orbit
planets
solar system

The Big QUESTION
Why do the planets move around the sun?

Think about the sun on a sunny day. You are probably aware that throughout the day, the sun seems to make a circle around Earth. The sun appears to start each day in the east, then to move overhead at midday. The sun appears to end the day by setting in the west. Observations like this led people in the past to incorrectly believe that the sun moved around Earth.

The Solar System

Today, we know that Earth and the other objects in space move, or revolve, around the sun. But why do objects in space revolve around the sun? What prevents them from simply floating away into space? The answer is gravity.

B4

You already are familiar with the force of gravity and its effects on objects on or near Earth. Every time you throw a ball into the air, gravity pulls the ball back to Earth. Gravity attracts people, animals, and other objects toward Earth.

Like Earth, the sun also exerts a force on objects. Because the sun has a very large mass, this force, called **gravitational pull** (grav′ə tā′shən əl pool), is very great and acts over long distances. This means that even though the sun is millions of kilometers from Earth, the gravitational pull between the sun and Earth keeps Earth traveling in a curved path around the sun. This path is called Earth's **orbit** around the sun.

Earth is not the only object in space that is affected by the sun's gravitational pull. Scientists know of eight other large objects that orbit the sun. These objects are called **planets,** after the Greek word for *wanderer.* We know now that these planets, including Earth, do not wander freely through space. They and many smaller objects orbit the sun, making up the **solar system.**

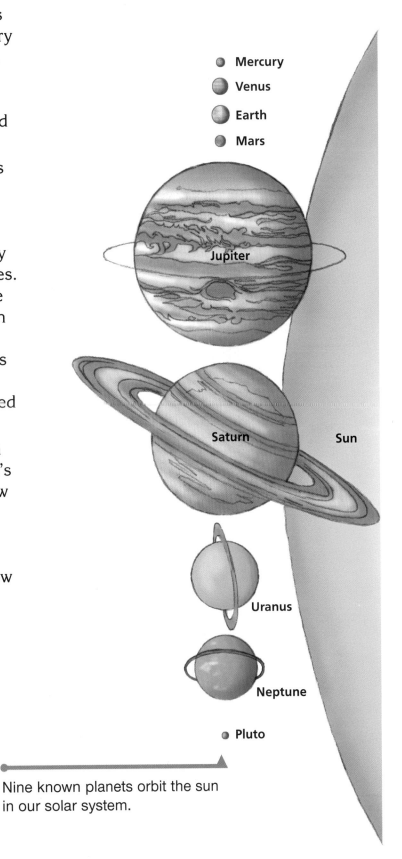

Nine known planets orbit the sun in our solar system.

The Planets

Inner Planets

The nine planets in our solar system can be put into two groups. The four inner planets are Mercury, Venus, Earth, and Mars. These planets are made mostly of rock and are smaller than the outer planets. Because these planets are closer to the sun, they are warmer than the other planets.

Mercury

Mercury is the planet closest to the sun. Mercury's surface is heavily cratered and looks like the moon. There is no atmosphere to protect it from the sun's rays. Mercury's temperature can reach 430 °C during its day, but during its night Mercury gets extremely cold.

Venus

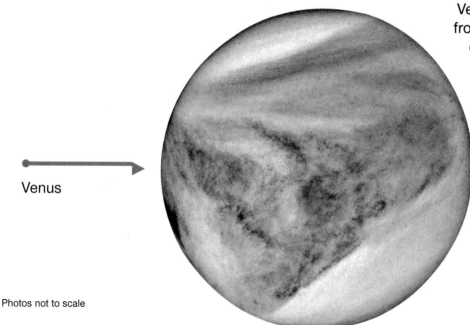

Venus, the second planet from the sun, comes the closest to Earth. Venus's atmosphere of sulfuric acid clouds and carbon dioxide traps the sun's heat, making it hotter than Mercury. The surface of Venus has a huge lava plain with mountains and giant volcanoes.

Photos not to scale

Earth

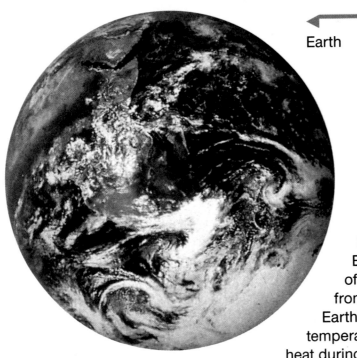

Earth is the third planet in the solar system. Earth's atmosphere contains oxygen, carbon dioxide, and, most importantly, water vapor. Earth's atmosphere traps some of the sun's heat and protects us from the sun's damaging rays. Earth's oceans also help keep the temperature moderate by absorbing heat during the summer and giving it off during the winter. Earth is the only planet with a large amount of liquid water.

Mars

Mars is the fourth planet from the sun. Mars is covered with iron-rich rocks and dust, so it appears to be reddish. Winds on Mars cause frequent dust storms. Scientists believe that Mars at one time may have had liquid water, but now it only has carbon dioxide ice in its polar ice caps.

Outer Planets

Four of the five outer planets are much larger than the inner planets. Pluto is the only one that is smaller. Because the outer planets are farther away from the sun, they are generally colder than the inner planets. The outer planets are known as "gas giants" because they are made mostly of gases like hydrogen, helium, methane, and ammonia.

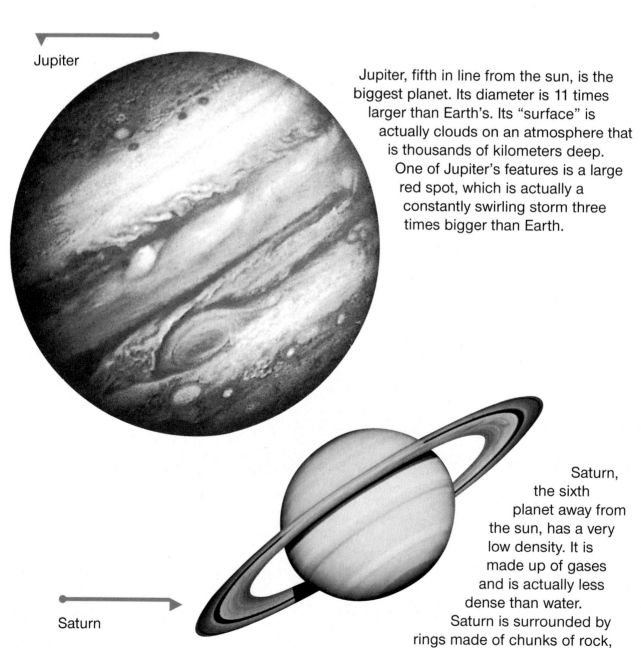

Jupiter

Jupiter, fifth in line from the sun, is the biggest planet. Its diameter is 11 times larger than Earth's. Its "surface" is actually clouds on an atmosphere that is thousands of kilometers deep. One of Jupiter's features is a large red spot, which is actually a constantly swirling storm three times bigger than Earth.

Saturn, the sixth planet away from the sun, has a very low density. It is made up of gases and is actually less dense than water. Saturn is surrounded by rings made of chunks of rock, dust, and ice. All of these materials spin around Saturn like millions of tiny moons.

Saturn

Photos not to scale

Uranus, the seventh planet from the sun, is about four times the diameter of Earth. It is so far from the sun that it is just barely visible to the eye. It looks greenish-blue because of the gases that make it up. Uranus is surrounded by eleven thin rings made of a black material.

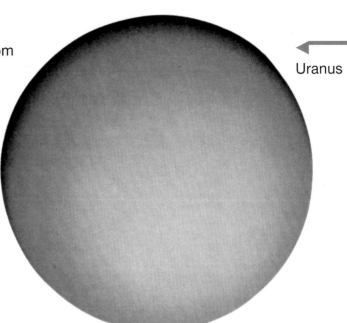
Uranus

Neptune

Neptune, the second-to-last planet in the solar system, is similar in makeup to Uranus. Because it is so distant, all we can see of it is a faint gray-green dot in a telescope. A visit by a space probe revealed to scientists that Neptune has dark clouds that come and go.

Pluto

Pluto, the farthest planet from the sun, is also the smallest. Pluto's surface is extremely cold. Pluto has an unusual orbit that sometimes takes it inside Neptune's orbit. Scientists predicted Pluto's existence before they actually discovered it with a telescope in 1930.

B9

Planet Years and Planet Days

A year is the time it takes Earth to make one complete orbit, or revolution, around the sun. Each planet takes a different amount of time to complete a revolution around the sun. The chart on the next page gives the length of time of each planet's orbit of the sun.

As every planet revolves around the sun, it also rotates on its axis. This is what creates day and night on the planets. The planets rotate at different rates. Some planets have very long days, and others have short days.

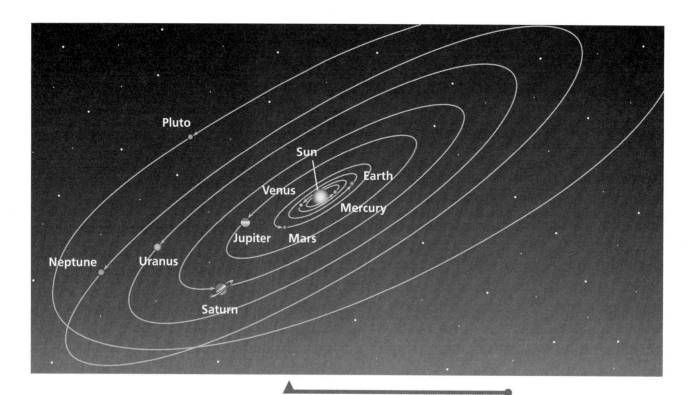

Trace Earth's orbit with your finger. Now find the orbits for Jupiter and Saturn. How are these orbits different from Earth's orbit?

Planet	Length of Day (one complete rotation in hours and days)	Length of year (one complete orbit in Earth days)
Mercury	59 days	88 days
Venus	243 days	225 days
Earth	24 hours	365 days
Mars	24 hours, 37 minutes	687 days
Jupiter	9 hours, 48 minutes	4332 days
Saturn	10 hours, 12 minutes	10,760 days
Uranus	17 hours, 54 minutes	30,681 days
Neptune	19 hours, 6 minutes	60,195 days
Pluto	6 days, 9½ hours	90,475 days

A year is the time it takes a planet to do one revolution around the sun. Do you see any pattern in the time it takes planets to travel once around the sun?

CHECKPOINT

1. What makes up the solar system?
2. Name one of the inner planets and one of the outer planets in our solar system.
3. How long does it take Earth to orbit the sun? How long does it take Mars?

 Why do the planets move around the sun?

ACTIVITY

Modeling the Solar System

Find Out
Do this activity to learn how to create a three-dimensional model of the solar system that represents the planets and their orbits around the sun.

Process Skills
Using Numbers
Constructing Models
Measuring
Communicating

What You Need

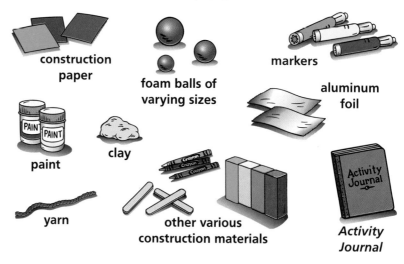

construction paper, foam balls of varying sizes, markers, aluminum foil, paint, clay, yarn, other various construction materials, Activity Journal

What to Do

1. Divide the class into ten groups. Write the name of each planet and the sun on slips of paper. Each group should pull one slip of paper to decide which group will make which model.

2. Use the information provided in this lesson, as well as other books, magazines, or references in your classroom to decide what characteristics your model should have. What color should it be? How big should it be? As a class, decide how big the entire model will be. Now your group should estimate how big to make your part of the model.

B12

3. Use the foam balls and any other materials you want to **make a model** of your planet. Remember to **compare** your model with others in the class as you go. This will help to make sure the planets are the right size.

4. When all groups have finished their individual planets and the sun, the class should put the whole model together. You could hang it from the ceiling or lay it out on a large table. Make sure that each planet's orbit around the sun is the correct size and shape.

5. Invite other classes to view your model, and **tell** them about your group's planet.

Conclusions

1. How is your group's part of the model different from the others?
2. How is your group's part similar to the others?

Asking New Questions

1. Pluto is sometimes not the farthest planet from the sun. How can this happen?
2. Which planets do you think are easiest to see from Earth?

SCIENTIFIC METHODS SELF CHECK

- ✔ Did I **compare** sizes and orbits with the other groups' so that our class model would correctly represent the solar system?
- ✔ Did I help make our planet or sun **model** look like the real one?
- ✔ Did I correctly **measure** our group's orbit?
- ✔ Did I **tell** others about our model?

LESSON 2

The Moon and Earth

Find Out
- What Earth's natural satellite is
- How Earth and the moon move
- How the moon affects Earth's tides

Vocabulary
satellite
tides

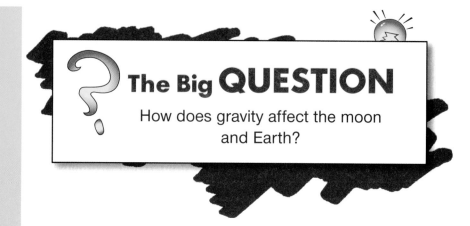

The Big QUESTION

How does gravity affect the moon and Earth?

How many times have you gazed up into the night sky and spotted a bright, full moon? At night, the moon is usually easy to spot. But have you ever wondered why the moon looks a little different each night?

Earth's Satellite

A **satellite** (sat′ ə līt) is an object that revolves around another object. Earth and the other eight planets are satellites of the sun. A moon is a natural satellite. Earth and Pluto only have one moon each. Mercury and Venus do not have moons. The other planets each have two or more moons.

The moon revolves around Earth because both Earth and the moon exert a gravitational pull on each other. If Earth's mass were smaller, the gravitational pull would be less and the moon's orbit would be different.

To see during the day, you need to remember that the moon is not a source of light. The moon only reflects the sun's light. The moonlight you see at night is really sunlight shining on the moon and reflecting back toward Earth.

Earth receives the sun's reflected light from the moon.

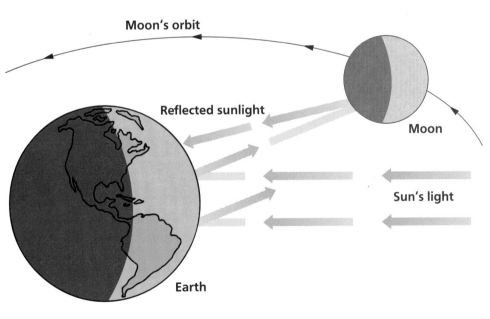

Movement of the Moon and Earth

Remember that Earth rotates on its axis as it revolves around the sun. The moon does much the same as it travels around Earth. The moon takes 29 1/2 days to complete one orbit around Earth. As the moon orbits Earth, it also rotates slowly on its axis. The moon takes about the same number of days to rotate as it does to revolve around Earth. Because these two motions take the same amount of time, the same side of the moon is always facing Earth.

You and a partner can demonstrate this using a large ball. You can act as Earth and stand in one place for this demonstration. Your partner should hold a ball in front of you. Then, have your partner walk around you, rotating the

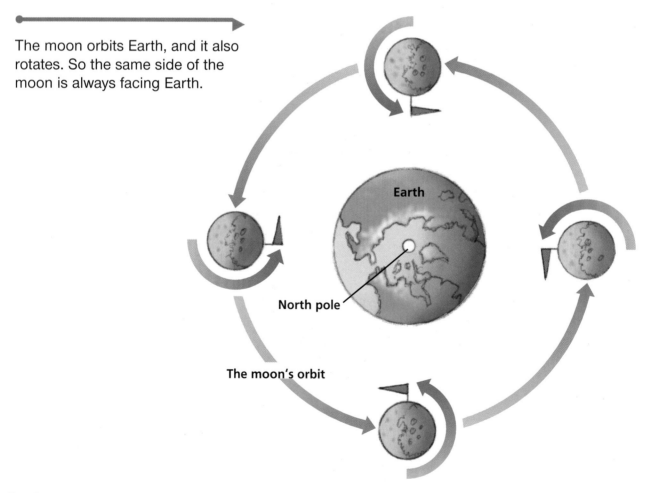

The moon orbits Earth, and it also rotates. So the same side of the moon is always facing Earth.

Earth

North pole

The moon's orbit

ball so the same side faces you. After your partner has walked completely around you, the ball will have rotated once and revolved once. You will see only one side of the ball, but your classmates will see all the sides.

Because the moon has less mass than Earth, its gravitational pull on objects is less. An object on the moon, like an astronaut, experiences a gravitational pull that is about one-sixth of the gravitational pull on Earth.

The moon's gravitational pull on Earth affects Earth in some interesting ways. For example, as Earth orbits the sun, the moon is also tugging at Earth. This causes Earth to "wobble" slightly as it travels on its orbit around the sun.

Why do you think astronauts can jump higher on the moon than they can on Earth?

In the Bay of Fundy, the difference between high and low tides can be as much as 21 m.

Tides and the Moon

An even more noticeable effect of the gravitational pull between the sun, Earth, and moon can be observed in Earth's oceans. Ocean **tides** are the rise and fall of the water's surface. Low tide is the best time to collect seashells at the beach, because there is a lot more uncovered land. When the tide is high, less land can be seen because the water level rises.

Why do tides occur? The gravitational pull between Earth and the moon causes the ocean water to bulge on the side of Earth facing the moon and on the side facing away from the moon. As Earth rotates, the water appears to rise and fall, resulting in high and low tides.

Twice a month when the sun, Earth, and moon all make a straight line, the added gravitational pull from the sun causes the water to bulge more. These extra-high tides are called spring tides because they appear to "spring up" very fast. When the sun, Earth, and moon are at right angles, the water does not bulge as much. These tides are called neap (nēp) tides.

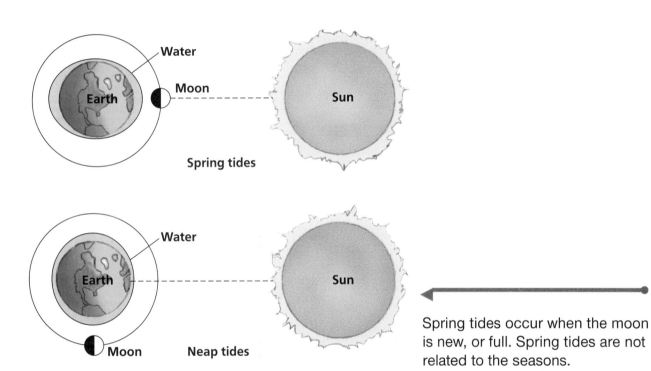

Spring tides occur when the moon is new, or full. Spring tides are not related to the seasons.

CHECKPOINT

1. What is Earth's only natural satellite?
2. Describe how Earth and the moon move.
3. What creates high and low tides?
4. How does gravity affect the moon and Earth?

ACTIVITY
Gravitational Pull

Find Out
Do this activity to see how Earth's gravity pulls on the moon.

Process Skills
Constructing Models
Observing
Predicting
Communicating

What You Need

a 30-cm piece of string

safety goggles

a metal washer

Activity Journal

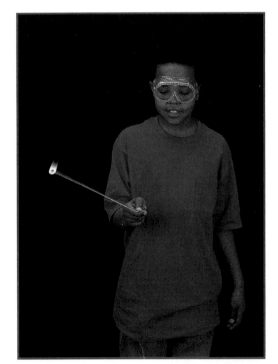

What to Do

1. Tie a washer to the end of the string.

    Put on your goggles.

2. Hold the string in your hand about 25 cm away from the washer. Then whirl it around.

    Stand well away from classmates, walls, and furniture.

3. **Observe** what happens. **Predict** what would happen if the string suddenly broke. Where would the washer go? **Record** your predictions.

CONCLUSIONS

1. What did the washer represent?
2. What did you represent when you whirled the washer around?
3. What did the string represent?
4. How do you think this is like the relationship between Earth and the moon?

ASKING NEW QUESTIONS

1. How would your life be different if Earth did not have a moon?
2. What do you think would happen to the moon if Earth's mass greatly increased?

SCIENTIFIC METHODS SELF CHECK

✔ Did I **model** Earth's gravitational pull on the moon by whirling the washer on the string?

✔ Did I **observe** how the washer moved?

✔ Did I **predict** what would happen if the string broke?

✔ Did I **record** my observations and predictions?

Space Exploration

Find Out
- How people explore space
- How telescopes help us see into space
- What tools people use in space

Vocabulary
telescopes
rockets
artificial satellites
space probes
space shuttles

The Big QUESTION

How can we explore space if gravity attracts objects toward Earth?

*H*ave you ever looked up into the night sky and wondered what was happening in outer space? Many other people have also looked at the same night sky. They have told stories and imagined what might be happening in space. Scientists have also tried to answer many questions by exploring space in a variety of ways.

Outer Space

Every day we put our five senses to use in many ways. We use them to gather data about the environment where we live. Without all of our senses, we would not be able to experience much of the world around us. We can adapt to the loss of one or two senses. But can you imagine all that you would miss if you could not see, hear, touch, taste, and smell?

Now suppose you wanted to use your senses to explore a place very far away, and you were unable to travel there. You would have to find a way to send your senses there without you. This is just what scientists face when trying to explore outer space.

Telescopes

One of the earliest ways people extended their sense of sight into space was through the use of telescopes. **Telescopes** are tools that collect light and bring it to a point of focus so the object that you are looking at can be magnified. This allows the person looking through the telescope to see distant objects more clearly. Telescopes allow people to study stars and planets in more detail.

The Hubble Space Telescope was released in space from the space shuttle *Discovery*. The telescope has sent back pictures to Earth of many distant stars.

Because sunlight makes the sky so bright on Earth, searching for stars and planets is generally easier to do at night. Clouds and bad weather also make it difficult to see the stars with telescopes. So large telescopes on Earth are often put in places far away from cities and their night lights—often high in mountains or in dry desert areas to reduce the problems from haze, smog, and clouds.

Tools in Space

Today, scientists also send many instruments into space to study the objects in space. Some tools scientists use are satellites, space probes, and space shuttles. Powerful rockets are needed to lift these tools into space and to overcome Earth's gravitational pull. **Rockets** are not tools used to study space; they are only the vehicles (vē′ hik əls) to carry the tools into space.

Rockets are moved by very hot gases. The great force of the gases leaving the base of the rocket creates the thrust that pushes the rocket upward. The tools that will be used to observe outer space or Earth are attached to the rocket.

Eugene and Carolyn Shoemaker use the Schmidt Telescope.

Satellites

You already know that a satellite is an object that orbits a larger body in space. As well as natural satellites, like the moon, there are **artificial satellites** built by humans.

Humans launch artificial satellites into space using rockets. The rocket accelerates the satellite far enough away from Earth so that the satellite will no longer be pulled back to Earth. When the satellite reaches a certain point in space, it can go into orbit around Earth.

Satellites are not only used to explore space. Some satellites are used for communication here on Earth. These satellites transmit radio and television programs, fax, phone, and computer data around the world. There are also weather satellites that record the atmosphere from above. Weather satellites help people forecast the weather and can show the location of dangerous storms.

Communication satellites bring television, fax, radio, and computer data to people around the world.

Weather satellites record the atmosphere from above. They help us forecast the weather.

B25

With information recorded by space probes, such as the Mariner 10, scientists can learn about distant planets without being there.

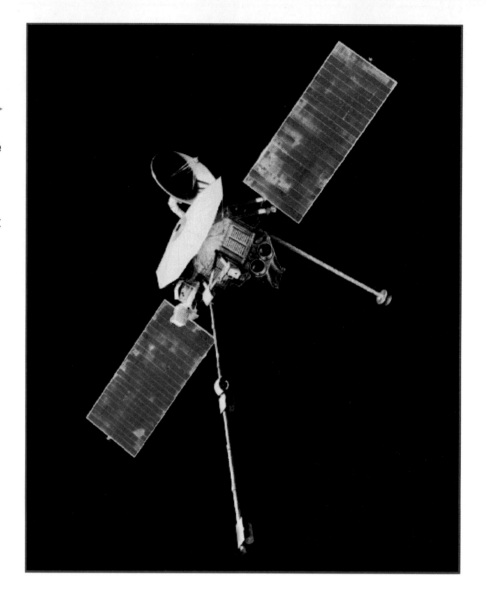

Space Probes

Like satellites, space probes are also launched by rockets in order to overcome Earth's gravitational pull and to travel far from Earth. **Space probes** travel close to distant planets or other objects in space. They do not orbit Earth or return; they are launched into space to help scientists learn about places too far away to send astronauts.

Space Shuttles

Another tool in space exploration is reusable **space shuttles.** A space shuttle depends on a rocket to launch it into space. Once above most of Earth's atmosphere, a shuttle remains in orbit around Earth. Crews aboard the shuttles do research, place satellites into orbit, and repair satellites already there. Some crews have even carried supplies to a space station.

After a mission is completed, space shuttles are able to be flown back to Earth and landed on a runway. Once the space shuttle is inspected, it can be used again.

Discovery is one space shuttle that orbited Earth. On October 29, 1998, *Discovery* was launched into space with a crew of seven astronauts. One of the crew was 77-year-old John Glenn. This was John Glenn's second trip into space. His first flight took place on February 20, 1962, when he became the first American to orbit Earth.

CHECKPOINT

1. How do people explore space?
2. How do telescopes help us see into space?
3. Name two tools people use in space.
 How can we explore space if gravity attracts objects toward Earth?

ACTIVITY

Making a Balloon Rocket

Find Out
Do this activity to see how rockets on space probes can carry an object to Mars.

Process Skills
Designing Investigations
Constructing Models
Communicating
Experimenting

What You Need

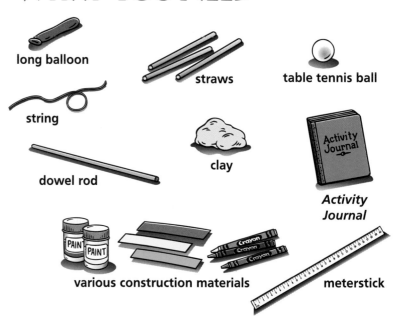

long balloon, straws, table tennis ball, string, clay, dowel rod, various construction materials, meterstick, Activity Journal

What to Do

1. Using the materials, design a balloon rocket that can carry the table tennis ball to Mars. Pretend that Mars is only 3 m from your launch site.
2. Draw the plan of your rocket design.
3. Then use the materials to build the rocket you have designed.

Safety! *To avoid spreading germs, only one student should blow up the balloon.*

4. **Mark** your launch site with an *X*. Draw a 50-cm circle 3 m from your launch site to represent Mars. **Test** fire your rocket.

   Safety! *Be sure everyone is out of the way before you launch your rocket.*

5. Make any changes you think will help get your rocket closer to Mars.

6. **Test** and **modify** your rocket three more times.

CONCLUSIONS

1. What caused your rocket to move?
2. What things had an effect on the flight of your rocket?
3. What changes did you make that helped your rocket fly better or fly worse?

ASKING NEW QUESTIONS

1. What did you change to make your rocket more accurate? How well did these changes work? If you were able to change your rocket again, how could you make it even more accurate?
2. Compared to scientists who build actual rockets, what advantages did you have in building yours? What disadvantages did you have?
3. Why do people build rockets?

> ### SCIENTIFIC METHODS SELF CHECK
> ✔ Did I **design** a rocket using the materials provided?
> ✔ Did I **test** and **modify** the rocket to help it land closer to the target?
> ✔ Did I **share** my ideas with the group?

Review

Reviewing Vocabulary and Concepts

Write the letter of the word below that best completes each sentence.

1. The sun's ___ keeps the planets revolving around the sun.
   - **a.** heat
   - **b.** gravitational pull
   - **c.** orbit
   - **d.** satellites

2. The path that a planet takes around the sun is called ___.
   - **a.** a revolution
   - **b.** an orbit
   - **c.** a meteorite
   - **d.** a space probe

3. The nine large objects that orbit the sun in our solar system are called ___.
   - **a.** planets
   - **b.** moons
   - **c.** satellites
   - **d.** comets

4. The planets and smaller objects that orbit the sun make up the ___.
   - **a.** galaxy
   - **b.** Milky Way
   - **c.** Earth
   - **d.** Solar System

5. The gravitational pull of the moon affects ___ on Earth.
   - **a.** tides
   - **b.** telescopes
   - **c.** televisions
   - **d.** space probes

Match the definition on the left to the word on the right.

6. one of the first instruments people used to help them see stars and other objects in space
7. used to launch tools for gathering information in space
8. orbits around Earth, taking pictures and gathering information
9. sent into deep space to gather information and explore faraway planets
10. a reusable spacecraft that carries people into space

- **a.** rocket
- **b.** space probe
- **c.** telescope
- **d.** space shuttle
- **e.** satellite

Understanding What You Learned

Review

1. What is gravity?
2. Why do planets orbit the sun and why does the moon orbit Earth?
3. Name some types of satellites that orbit around Earth. Tell whether they are natural or artificial.
4. How does the moon affect Earth's tides?
5. What are some of the tools we use in space exploration?

Applying What You Learned

1. How are the inner and outer planets different?
2. Why do we see the moon at night?
3. How can scientists break away from Earth's gravitational pull?
4. Why do scientists send space probes into deep space?

5. What are some ways gravity affects the solar system?

For Your Portfolio

Use any common objects in your desk or around the classroom to model how the parts of the solar system move. If an object has an orbit, describe what it goes around. If an object rotates, show that.

CHAPTER 2

Weather

Most of the time, people don't think much about the weather. It's usually when the weather is bad that we pay attention. Actually, the weather that happens in cities and rural areas affects many other things in our lives.

Food prices go up and down depending on whether or not farmers have had good weather. Mail and package deliveries slow down when rain, fog, or snow make truck or plane travel difficult. Telephone calls and e-mail messages stop when storms knock down communication lines. Whether you notice or not, air, water, and the sun are interacting to make Earth's weather patterns. The weather affects all of us.

The Big IDEA

Weather conditions change and can be observed and measured.

CHAPTER SCIENCE INVESTIGATION

Make a weather station and predict the weather. Find out how in your *Activity Journal.*

B33

Lesson 1

Air and Wind

Find Out
- What air is
- What causes wind
- How wind can be used

Vocabulary
air
mass
air pressure
wind

The Big QUESTION

How does air affect the weather?

Air is invisible, but it has a lot of power. The dentist's drill uses air power. Air bags are used in cars to protect drivers and passengers from contact with hard surfaces in case of an accident. Parachutists use air to their advantage as they glide gently toward land. Air is also a big factor in wind and weather.

Properties of Air

People and other animals need air to survive because air contains oxygen and other necessary gases. You could live without food or water for several days, but you would live only minutes without the gases in air.

Air is invisible, but you can see that it takes up space when a balloon expands as you blow it up. You can see air rippling the grass and whipping lakes and oceans into waves. You can't smell air, but the odors you can smell reach your nose because air carries them there.

But what *is* air? **Air** is about one-fifth oxygen and four-fifths nitrogen. It also has small amounts of other gases, including water vapor and carbon dioxide.

Air's most important function for human beings and other animals is that it allows plants and animals to exchange the oxygen and carbon dioxide that all living things need to survive.

You know that air takes up space. You can't fit a blown-up balloon in your pocket the way you can fit one that is not full of air. The air contains many very small molecules, so it takes up space in the balloon. Air also has mass. **Mass** is the amount of matter that something contains.

Because air has mass, it also exerts pressure. On the surface of Earth, you are at the bottom of a layer of air that is hundreds of kilometers deep. The weight of air pushing down on Earth's surface helps create **air pressure.**

Air and Gravity

If we measure air pressure high up in the atmosphere, it is lower than the air pressure at Earth's surface. Earth's gravitational pull holds most of the air close to the surface. If you go up to higher altitudes—for example, when you fly up in an airplane—there is less air above you, so you feel less air pressure.

You have felt the change in air pressure if you have taken an elevator to the top of a tall building or flown in an airplane. Now that you

Having fun with a giant bubble. It's filled with air!

know air pressure is lower at high altitudes, can you explain why your ears pop on elevators or in airplanes? Because the air pressure pushing on your body has decreased, the pressure inside your body pushes out, causing a slight "pop" sound in your ears.

Wind

About 500 years ago, European sailors began to understand how to find winds that usually blew in the same direction. They gave names to these steady winds and marked them on charts, much as we mark highways on maps today. What they learned has helped us understand what causes wind.

Wind is formed when the sun heats both land and water. As they warm up, so does the air above them. But land usually warms much faster than water, and so does the air above it. As the air over the land gets warmer, its molecules move farther apart. This causes the air to spread out and become less dense. As the air becomes less dense, it rises. The cooler air over water has its molecules closer together, so it is more dense. So there is more air over the cooler water. This makes the air pressure higher over the water than over the land. Air moves from higher-pressure areas to lower-pressure areas. We call this air movement **wind.**

The cooler air is like the air let out of a balloon. It moves to an area of lower pressure. What do you think happens at night when the land becomes cooler than the water?

Why Breezes Blow

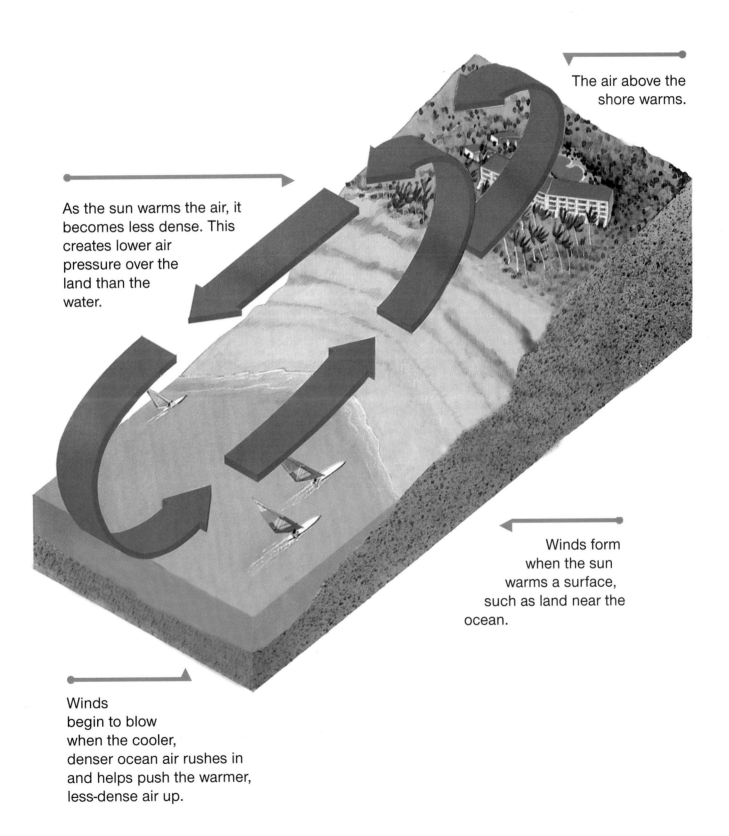

The air above the shore warms.

As the sun warms the air, it becomes less dense. This creates lower air pressure over the land than the water.

Winds form when the sun warms a surface, such as land near the ocean.

Winds begin to blow when the cooler, denser ocean air rushes in and helps push the warmer, less-dense air up.

Using the Wind

From a soft, gentle breeze to a powerful hurricane, the wind is a part of everyday life. The wind can be a valuable resource when it is used to make jobs easier. As fuel supplies, such as coal, oil, and natural gas, become more scarce and expensive, people are again turning to an old source of energy—wind.

A condor uses currents in the wind for flying.

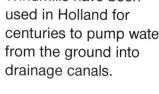

Windmills have been used in Holland for centuries to pump water from the ground into drainage canals.

Wind can be a good source of energy for several reasons. First, energy from the wind does not cost much as long as you have access to it. Second, it is clean energy because it does not cause pollution. In areas where strong winds blow regularly, as in some areas of California, the wind has become a reliable source of energy.

Plants and animals also need the wind. Birds, bats, insects, and even flying squirrels all use the wind for travel. The wind carries the scents of animals so they can avoid or meet with each other. The wind also spreads plant seeds for pollination.

People have tried to control the wind for many reasons. Some people have used wind for transportation. Others use it to help make jobs easier. There are also some sports that use wind. Windsurfing and sailing are just two sports. Can you think of others?

Windsurfers rely on the wind.

CHECKPOINT

1. What is air?
2. What causes wind?
3. Name two ways wind can be used.
 How does air affect the weather?

ACTIVITY

Making Magic with Air

Find Out
Do this activity to see how the properties of air give you evidence that air is all around you.

Process Skills
Predicting
Communicating
Observing
Experimenting
Inferring

What You Need

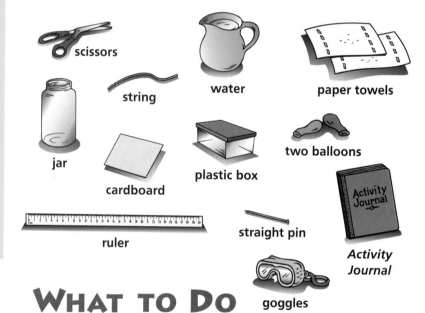

scissors, string, water, paper towels, jar, cardboard, plastic box, two balloons, ruler, straight pin, goggles, Activity Journal

What to Do

Part A

1. Fill the plastic box half full with water.
2. Crumple a paper towel and place it in the jar.
3. Hold the jar with the open end down and push it down to the bottom of the plastic box without tilting it. **Record** your observations.

Part B

1. Cut a square from the cardboard that is larger than the diameter of the plastic jar.
2. Fill the jar half full of water. Wet the cardboard and place it over the mouth of the jar.

3. **Predict** what will happen when the jar is turned upside down. **Write** down your prediction.
4. Flip the jar over as you hold the cardboard in place. Take your hand off the cardboard. What happens?

Part C

1. Tie an inflated balloon to each end of a ruler. Tie a string to the middle of the ruler. Hold the string so that the ruler is suspended. Move the balloons in or out to balance the ruler.
2. **Predict** what will happen if one balloon is popped. **Write** down your prediction.
3. Using a straight pin, pop one balloon.

    Be careful when using the straight pin. Be sure to wear goggles when popping the balloons.

CONCLUSIONS

1. In Part A, why doesn't water fill up the glass and get the paper wet?
2. How can you explain what happened in Part B?
3. In Part C, which end of the ruler dropped? What property of air caused this to happen?

ASKING NEW QUESTIONS

1. Which property of air causes your ears to pop?
2. Which property of air helps you keep a bicycle tire inflated?

SCIENTIFIC METHODS SELF CHECK

✔ Did I **predict** the effects of air?

✔ Did I **test** for and **observe** the properties of air?

✔ Did I **infer** why my tests had the results they did?

Water Cycle and Clouds

Find Out
- What the water cycle is
- What different types of clouds can form
- What rain is

Vocabulary
water cycle
water vapor
evaporation
condensation
precipitation
rain shadow

The Big QUESTION
How are clouds related to the water cycle?

*Y*ou can see water in lakes, rivers, oceans, and even frozen water in icebergs and glaciers. After it rains, you might see water in puddles on the street. But a day or two after a rainstorm, the water in puddles has disappeared. Where does this water go? Where does the water come from? The answer to both questions is that water goes around and around in a cycle.

The Water Cycle

Sometimes, in the early morning, if you walk barefoot, you may feel water in the grass. You may also find drops of water on leaves in the garden. Where does this water come from? Later in the day, the water will be gone. Where did it go? After it rains, water collects in streams that flow into rivers, which flow into lakes or oceans. The oceans are very large, but why, after centuries of collecting water, don't they overflow? This pattern of the appearance and disappearance of water is called the **water cycle.**

Remember that air contains oxygen, nitrogen, and small amounts of several other gases. One of these other gases is **water vapor.** Water has the ability to change from liquid to gas by a process called **evaporation** (i va′ pə rā′ shən). This is what happens when water seems to disappear. Liquid water molecules start to move faster and farther apart as they heat up. The molecules leave the surface in the form of gas. Liquids evaporate faster when they are heated. When water vapor cools, it turns into a liquid. Water vapor turning into a liquid is called **condensation** (kon′dən sā′ shən). Water can also form a solid by freezing as ice crystals.

When water evaporates and turns into gas, it rises into the atmosphere. Up there it cools, forming droplets of water which form clouds. The water falls again as rain, snow, hail, or sleet when the concentration of droplets becomes too great. These droplets of water are called **precipitation** (prē si′ pə tā′ shən). Water vapor that cools and condenses on the ground or on plants is what makes dew on the grass in the morning.

You have been in a cloud-making chamber if you have ever taken a warm bath on a cold day. Once you've filled the tub with warm water, what do you see rising out of the water? What forms on the cold surfaces in the room, such as the mirror and the window?

When water vapor cools, it turns into a liquid.

Oceans And The Water Cycle

The bathtub example shows how water travels back to the air from the oceans. Ocean water is warmed by the sun, evaporates, and rises into the air. High in the atmosphere, the air temperature is cooler, like the air temperature in your bathroom. Water vapor in the air condenses into clouds which can produce rain or snow, just as the water vapor in the bathroom condenses on the mirror and the window.

The water cycle is the repeating cycle of evaporation, condensation, and precipitation that occurs on Earth.

In Earth's water cycle, the sun is the energy source that causes evaporation.

Water condenses to form clouds of water droplets or ice particles suspended in the air.

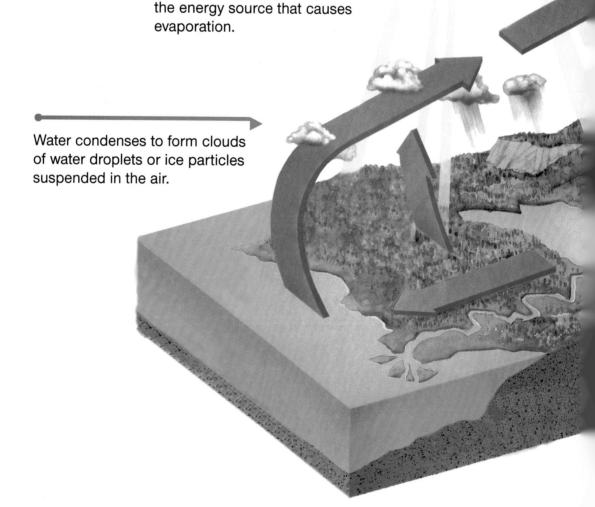

Some water runs into streams and some seeps into the ground. Some water is absorbed by plants and is returned to the atmosphere through evaporation and transpiration. Transpiration (trans′ pə rā′ shən) is the release of water vapor from plants. Some water runs from rivers into the oceans. The oceans do not overflow because water evaporates from them.

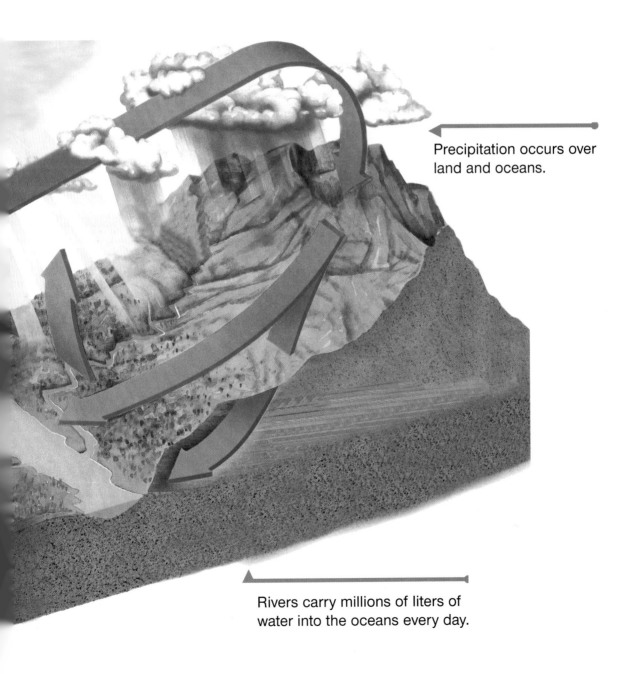

Precipitation occurs over land and oceans.

Rivers carry millions of liters of water into the oceans every day.

Types of Clouds

Have you ever been out on a foggy morning? Did it seem as if you were in a cloud? Actually, you were. Fog is a cloud that has formed close to the ground. Fog is formed the same way that all other clouds are formed.

There are three main types of clouds. Each type occurs at different heights and has different forms. You can predict the precipitation for tomorrow by observing the clouds today.

Cumulus (kyōō′ myə ləs) clouds are large, thick, and puffy. They often look white with gray centers. Cumulus clouds are usually flat on the bottom and pile up to look like a dome. These clouds often form on hot summer days. Sometimes they turn into thunderclouds or thunderheads. When this happens, rain or thunderstorms may occur.

Cirrus (sir′ əs) clouds are made of tiny ice crystals. They are thin and white with feathery edges. They are the highest clouds in the sky. You can see them when the weather is cool and dry.

Stratus (strā′ təs) clouds form close to Earth's surface. They can hold a lot of moisture, and they spread across the sky in flat, gray layers.

Getting Rain

Water is necessary for plants and animals to live and grow. Some areas of the world have lots of water in lakes, rivers, and oceans. Other areas get little rainfall, such as areas that are separated from an ocean by high mountains. Places that get less than 25 cm of rain a year are called deserts. Very few plants grow in deserts because there is not enough water to keep them alive.

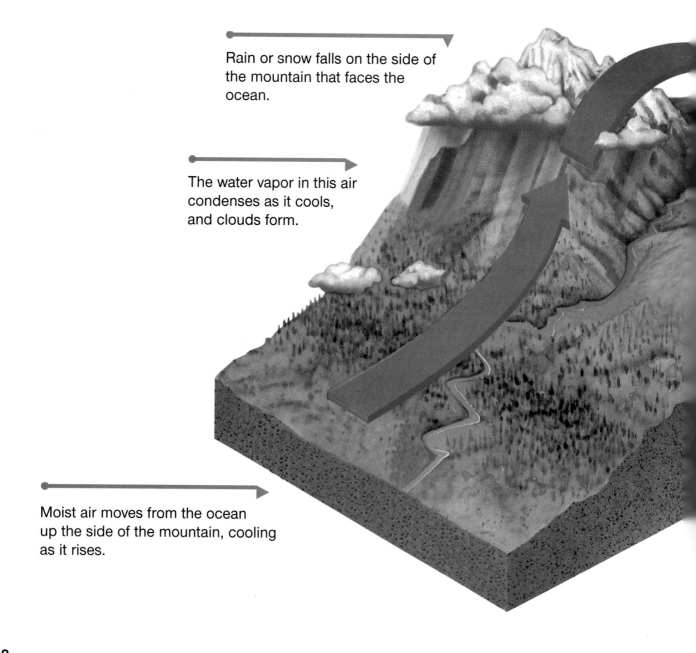

Rain or snow falls on the side of the mountain that faces the ocean.

The water vapor in this air condenses as it cools, and clouds form.

Moist air moves from the ocean up the side of the mountain, cooling as it rises.

By knowing how the water cycle works, you can see why some areas get less rain than other areas. The land on the side of a mountain away from the ocean gets little rain. This area is called a **rain shadow.** Most North American deserts are formed by rain shadows.

Understanding the water cycle will help you understand the weather and how it changes. The water cycle is one of the cycles that form our weather.

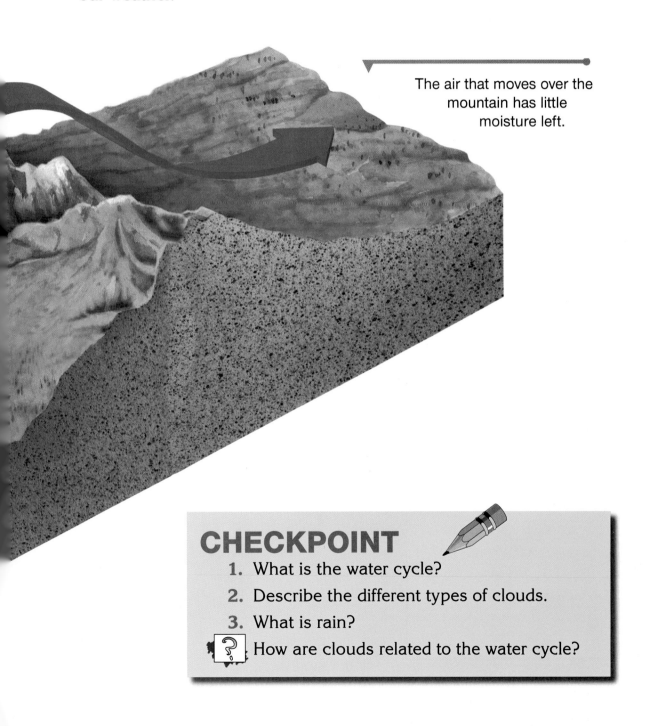

The air that moves over the mountain has little moisture left.

CHECKPOINT

1. What is the water cycle?
2. Describe the different types of clouds.
3. What is rain?
4. How are clouds related to the water cycle?

ACTIVITY

Making a Cloud

Find Out

Do this activity to see how condensation makes a cloud.

Process Skills

Observing
Communicating
Inferring
Predicting

What You Need

jar or clear plastic cup

water

ice cubes

watch or clock

paper towels

Activity Journal

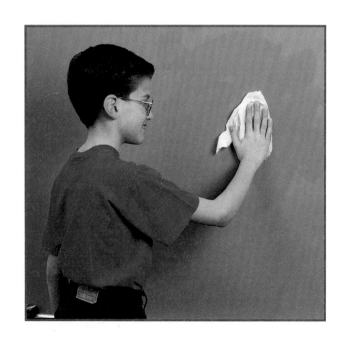

What to Do

1. Wet a piece of paper towel and wipe some water on the chalkboard. **Observe** the water. Then, **record** your observations.

2. Fill the jar halfway with water. **Observe** the outside of the jar and **describe** how it looks.

3. Now, add enough ice to fill the jar. Continue to **observe** the jar. Every 10 minutes, **record** changes that occur on the outside of the jar. Do this for 30 minutes.

Conclusions

1. What did you **observe** when you wiped the chalkboard with the wet towel?
2. What did you **observe** after you poured the water into the jar?
3. What did you **observe** after you added ice to the jar?

Asking New Questions

1. In the jar of ice and water, where did the water on the outside of the jar come from?
2. **Predict** what would happen to the water around the bottom of the jar if you let it sit for two days.

SCIENTIFIC METHODS SELF CHECK

✔ Did I **observe** changes to the water?

✔ Did I **record** my observations?

✔ Did I **observe** the water over the 30 minutes?

✔ Did I try to **explain** why the water disappeared and appeared?

Severe Weather

Find Out

- What an air mass is
- What happens when a warm front moves in
- Where hurricanes and tornadoes form

Vocabulary

air mass
front
thunderhead
cold front
warm front
hurricane
tornado

The Big QUESTION

What causes severe weather?

You have probably spent some time listening to the weather reports on TV or the radio about an upcoming storm. Scientists study the weather and prepare these reports to help you make your plans for the day. Will school be canceled because of the stormy weather? To forecast the weather, scientists study wind patterns. They use satellites and radar. Sometimes the forecasters are right, and sometimes they are wrong.

Air Masses and Fronts

The four factors that affect weather are temperature, air pressure, moisture, and wind. When weather forecasters predict weather, they use technology such as satellites and radar to gather data on these factors. They also track the movement of air masses. An **air mass** is a large body of air with nearly the same temperature and moisture throughout. An air mass may form over either land or water. It will take on the properties of the water or land it forms over: cool and more dense over water, and warm and less dense over land.

Air masses are usually slow to mix with other air masses. When a warm and a cold air mass meet, the warm mass rises up and over the cold mass. Or the cold mass can slide under the warm mass.

Air masses move as the wind blows. When they move, they bump into or push against other air masses. The place where two different air masses meet is called a **front.** Most storms form along fronts.

A satellite image of a severe storm in the Bering Sea

Storms

Storms develop at fronts because of rapid changes in temperature, humidity, and pressure. A thunderstorm forms when warm, humid air along a cold front is suddenly pushed up. As the warm air rises, it cools, and the water vapor condenses and forms a dark thundercloud. As more air is pushed up, the cloud grows taller, forming a **thunderhead.**

Thunderheads can produce strong lightning, thunder, and heavy rains. Most thunderstorms in the United States occur from May through September. Thunderstorms can be very dangerous.

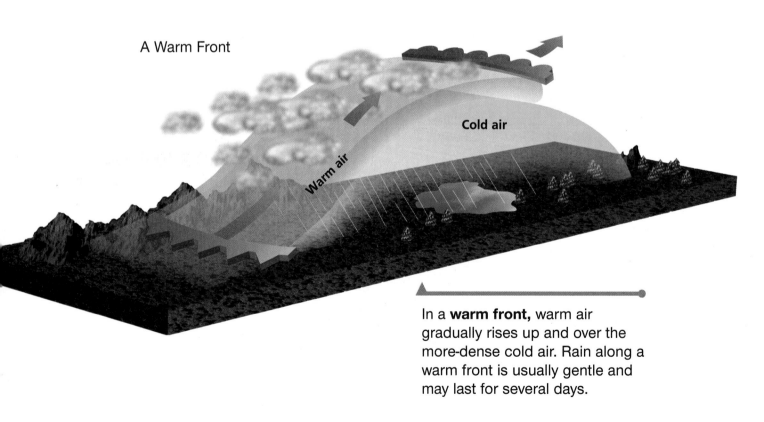

A Warm Front

In a **warm front,** warm air gradually rises up and over the more-dense cold air. Rain along a warm front is usually gentle and may last for several days.

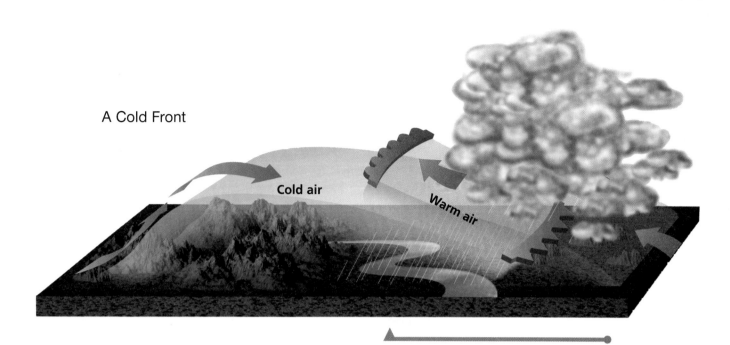

A Cold Front

A **cold front** produces fast-moving storms and strong winds. The dense cold mass pushes under the less-dense warm mass. The warm, moist air is forced to rise quickly. Rain at cold fronts is often heavy, but it doesn't last very long. Weather following a cold front is usually clear, dry, and colder.

Sometimes precipitation falls to Earth as hail. Hail forms when water droplets in a thunderhead are quickly moved up and down through layers of cold air. The drops freeze into ice crystals and grow large enough to fall to the ground as lumps of ice. Often, it is warmer on the ground. But remember, hail is formed high in the sky where the air is colder.

Precipitation can also fall as snow crystals. Snow forms in the cold parts of the clouds. Snow can fall to the ground like a thick carpet when the winds are light. When heavy snow is joined with high winds, we call the storm a blizzard.

Hurricanes and Tornadoes

A **hurricane** (hûr′ ə kān) is a violent tropical storm that forms over a warm ocean. Hurricanes form from warm, moist air rising from the ocean and can cover many kilometers. Usually, hurricanes are spotted while they are still far out at sea, days before they reach land. People who live along the shore are warned so they can get ready for the wind, rain, and waves that are coming.

Hurricanes, known as typhoons in Asia, do their damage to islands or coastal lands. The strong winds and rains can destroy buildings and pull trees out of the ground. Sometimes getting ready for a hurricane means boarding up windows and bringing in the animals. Sometimes it means getting out of the area.

Hurricanes form over warm ocean water.

Hurricanes are named in alphabetical order as they develop, with male and female names alternating. During hurricane seasons there may be several hurricanes at one time. The names help people keep track of which one they're waiting for!

A **tornado** (tor nā′ dō), or cyclone, is also a storm with a powerful circular wind but the wind creates a funnel that is only meters across. It starts over land in thick storm clouds, not over water. A tornado forms when air masses of different temperatures collide. When the swirling funnel dips down to the ground, the strong wind can turn over buildings and pick up cars. After a tornado passes through, large pieces of debris can be found kilometers away from where they had been.

A tornado is a storm with a powerful circular wind that creates a funnel.

Tornadoes are smaller than hurricanes, but they often have greater wind speeds. Forecasters can only say that weather conditions are right for tornadoes to form. They may have to wait until one is spotted or picked up by radar to warn people in the area.

Hurricanes last longer and cover a much wider path, but they are more predictable than tornadoes. With days of notice, people can make plans, but the news everyone waits for is that the hurricane has gone back out to sea or has become too weak to cause damage.

CHECKPOINT

1. What is an air mass?
2. What happens when a warm front moves in?
3. Where do hurricanes and tornadoes form?
 What causes severe weather?

ACTIVITY

Making a Barometer

Find Out
Do this activity to learn how to make a weather instrument called a barometer to measure changes in air pressure.

Process Skills
Measuring
Controlling Variables
Observing
Communicating
Interpreting Data
Predicting

What You Need

What to Do

1. Cut the neck off of the balloon. Stretch the balloon tightly over the open top of the can. Hold the balloon in place with a rubber band.

   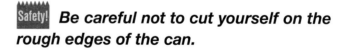 *Be careful not to cut yourself on the rough edges of the can.*

2. Tape the straw on the balloon so that one-third of the straw sticks out beyond the edge of the can.

3. Hold the index card lengthwise and write "Higher" at the top and "Lower" near the bottom. Fold the card lengthwise about 1 cm from the side. Tape the folded part of the

card just behind the straw that is hanging over the edge of the can. Use the straw as a guide to mark the card with a black line and **write** "Day 1" next to the line.

4. Set up the thermometer near the barometer.

5. At the same time each day, mark the position of the straw on the card with a black line and write "Day 2," "Day 3," and "Day 4." Each day, mark on a bar graph whether the pressure is rising or falling. **Record** the air temperature, sky color, and clouds. **Compare** your observations with the findings from the other groups and with a newspaper or television weather report.

CONCLUSIONS

1. What happened to make your barometer record a high pressure on the card?
2. What type of weather did you see on days with high air pressure? What type of weather did you see on days with low air pressure?
3. On the basis of your observations, can you **predict** what weather conditions will be on the fifth day?

ASKING NEW QUESTIONS

1. What connection did you notice between air pressure and temperature?
2. What type of reading would you expect from your barometer if warm air were to move into your area?

SCIENTIFIC METHODS SELF CHECK

✔ Did I **observe** and **measure** changes in air pressure?

✔ Did I make my observations at the same time every day?

✔ Did I **record** my observations?

✔ Did I try to **explain** how temperature, weather conditions, and air pressure are all related?

Review

Reviewing Vocabulary and Concepts

Write the letter of the answer that completes each sentence.

1. ___ is one-fifth oxygen and four-fifths nitrogen.
   - a. Helium
   - b. Air
   - c. A cloud
   - d. Gas

2. The weight of air pushing down on Earth's surface is ___.
   - a. gravity
   - b. force
   - c. air pressure
   - d. radiation

3. The process when water turns from liquid to gas is ___.
   - a. evaporation
   - b. photosynthesis
   - c. flotation
   - d. melting

4. A large body of air with nearly the same temperature and moisture throughout is ___.
   - a. an air mass
   - b. a cloud
   - c. a storm
   - d. a bubble

5. In ___, warm air slowly rises up and over the more-dense cold air.
   - a. a hot-air balloon
   - b. a warm front
   - c. a thunderhead
   - d. a rain cloud

Match each definition on the left with the correct word.

6. the amount of matter that something contains
7. air moving from higher-pressure to lower-pressure areas
8. pattern of appearance and disappearance of water
9. process of water vapor turning to liquid
10. a violent storm that forms over an ocean

- a. wind
- b. hurricane
- c. condensation
- d. mass
- e. water cycle

B60

Understanding What You Learned

1. What is air?
2. What causes wind?
3. Name the different types of clouds and describe them.
4. What is a front?
5. Where do hurricanes and tornadoes form?

Applying What You Learned

1. How does air affect weather?
2. How do clouds form?
3. Describe the difference between warm and cold fronts.
4. Often, when hail falls, the temperature is warm. How can this be?
5. How do weather conditions change, and how can these changes be observed and measured?

For Your Portfolio

Have you ever been through a really bad storm? Or have you watched one on the news? Write a story based on that experience. How did you feel and what did you see when you knew the storm was coming? During the storm? Afterward? How did the people around you react to the storm?

CHAPTER 3
Shaping and Reshaping Earth

Waves, wind, water, and ice shape Earth's surface. Slow processes, like erosion, cause some changes on Earth. Running water erodes landforms, reshaping the land by removing it from some places and depositing it as pebbles, sand, silt, and mud in other places. Rapid processes, like landslides, volcanic eruptions, and earthquakes, cause other changes.

The effects of some changes are obvious right away, like the piles of mud left by a landslide. Other changes take many, many years before their effects can be seen, like the changes on a gravestone or in the chemical makeup of soil. Nevertheless, Earth's surface is constantly changing.

The Big IDEA

Changes in Earth's surface occur when rocks are acted upon by various forces.

CHAPTER SCIENCE INVESTIGATION

Learn how weathering and erosion affect soil. Find out how in your *Activity Journal.*

Lesson 1

Weathering

Find Out
- How weathering can change rocks
- How water affects weathering
- What other factors affect weathering

Vocabulary
physical weathering
chemical weathering
freezing
thawing

The Big QUESTION
What is weathering?

*N*othing stays the same forever. The smooth edges and broken pieces on grave markers and rock structures like Mount Rushmore show how the forces of nature are continually changing structures, rocks, and landscapes.

Changes in Rocks

You can find rocks everywhere. They can show up in your garden or on your school playground. Rocks are hard and can be cut or broken into many useful shapes; therefore, people use rocks to build things. They might make up the sidewalk in front of your home or the parking lot where your school bus waits during the day. Some buildings in your town are probably made from different kinds of rocks. Even the cemetery in your area may use rocks to mark graves so we can remember where people are buried. Humans have used different types of rocks for building for as long as we know. Rocks may be hard and durable, but they do not stay the same forever.

Many rocks change color, composition, hardness, or form because of weathering. Weathering is the breaking down of rocks into smaller pieces by natural processes. The worn-down appearance of Mount Rushmore shows the results of weathering over years. If you go to a cemetery and look at very old grave markers, you can also observe the worn-down appearance caused by years of weathering.

The breakdown of rock involves two types of weathering that work together. These two types are physical weathering and chemical weathering. **Physical weathering** changes rocks through the forces of wind, water, or ice. Rocks that have been physically weathered provide some of the material found in soils. Wind can blow sand against other rocks. This constant hitting of sand against the rocks can break off small pieces of the rocks, changing their shapes. Water can cause rocks in a stream to bump together, breaking off small pieces. Even ice can change the shape of rocks by breaking them apart.

But rocks can also be changed by things other than wind, water, or ice. **Chemical weathering** is when substances that make up the rock are changed. This occurs when the acids in rain, air, or soil dissolve the minerals in rocks. These changes weaken a rock's structure so that it is more easily broken down by physical weathering. Elements in rain, air, soil, and plants can cause chemical weathering.

Mount Rushmore is made of rock, and shows signs of weathering.

Water and Weathering

Freezing and thawing can rapidly weather rocks.

Water plays an important role in the weathering process. **Freezing,** when a liquid changes to a solid by cooling, causes physical weathering. Water can seep into cracks and freeze when the temperature drops. As the water freezes, it expands and pushes the rocks apart.

When temperatures warm, thawing occurs. **Thawing** happens when something that is frozen solid changes to a liquid by warming. When ice thaws, it contracts, and the water goes down deeper into the cracks between rocks. If the temperature drops enough, the water will freeze again, causing the rocks to crack even more.

Freezing and thawing can weather the rocks on mountains very quickly because of the warm temperatures during the day and cold temperatures at night.

When minerals dissolve, caves can form. The minerals are carried away, leaving a large space.

Water can also be involved in chemical weathering. Sometimes the minerals in rocks can mix with water and dissolve. The dissolved minerals can then be carried away with the liquid. Chemical weathering may be a much slower process than the freezing and thawing of physical weathering.

How fast physical and chemical weathering occur depends on the size and composition of the rocks being weathered. However, pollution caused by people can speed up the chemical weathering process. Cars and factories pollute the air and water with substances that can cause chemical weathering. This is a problem for buildings, statues, and other structures made from rocks and minerals.

Repairing the damage caused by weathering and pollution is expensive. In the United States, cleaning and repairing statues and buildings harmed by chemical weathering from pollution costs several billion dollars every year.

It cost millions of dollars to clean and repair the Statue of Liberty after years of weathering.

Factors Influencing Weathering

Material

Many other factors affect the rate at which rocks weather. Different types of rocks weather at different rates. Think about the grave markers in old cemeteries. Markers set out at the same time can weather at different rates because some markers are made of limestone and others are made of granite. Limestone is a sedimentary rock and is quite soft. Granite, on the other hand, is an igneous rock and is very hard. Because of the difference in their hardness and composition, limestone will weather faster than granite.

Condition

You know that the type of rock affects the way it weathers, but the condition of the rock matters as well. Cracked rocks and rocks with holes will weather faster than rocks without these features.

Filling in cracks and potholes can reduce weathering on roads and highways.

Things made from rocks can have increased weathering because of cracks and holes. You may have seen potholes in roads. Weathering causes these potholes to form and deepen. Water can seep into the cracks in the road and freeze, causing larger cracks or holes to form. Road crews attempt to slow this weathering by filling in the cracks or holes. Sealant can also be placed on the cracks in driveways and parking lots to stop water from forming larger cracks and holes.

Chemical weathering happens faster in hot, humid climates.

Climate

Climate also affects the rate of weathering in rocks. Cold climates produce rapid physical weathering from freezing and thawing. Rocks in areas with hot, humid climates have great amounts of chemical weathering. High temperatures, large amounts of rainfall, and large numbers of plants result in more rapid chemical weathering. Water is a major factor in chemical weathering, so rocks in wet climates weather quickly. In desert areas, the lack of water slows down the process.

CHECKPOINT

1. How do rocks change from weathering?
2. How does water affect weathering?
3. Name three factors that influence how rocks are weathered.

 What is weathering?

ACTIVITY

Ice Power

Find Out
Do this activity to see how freezing water is involved in weathering.

Process Skills
Measuring
Controlling Variables
Predicting
Communicating
Observing

WHAT YOU NEED

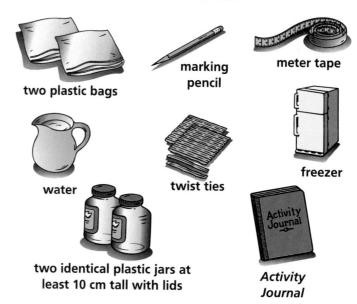

two plastic bags
marking pencil
meter tape
water
twist ties
freezer
two identical plastic jars at least 10 cm tall with lids
Activity Journal

WHAT TO DO

1. Beginning with 0 at the bottom of the jar, make marks at 1-cm intervals up the side of each jar. Fill each jar with water exactly to the 10-cm mark. Place the lid on each jar.

2. Place jar 1 inside a plastic bag and tie it shut. Place the jar in the freezer and leave it there overnight.

3. Place jar 2 inside a plastic bag and tie it shut. Leave it on a table overnight.

4. The next day, compare the appearance of the water inside each jar. Compare and measure the water levels inside each jar.
5. Make a bar graph that shows the levels in each jar.

CONCLUSIONS

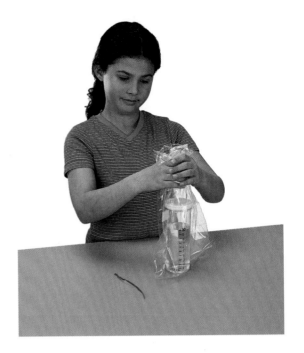

1. What happened to the water in the freezer?
2. Which jar contained the greater volume of matter?
3. What kind of weathering have you simulated?

ASKING NEW QUESTIONS

1. Predict what would happen if you fill a jar to the top with water and put the closed jar outside on a very cold night.
2. If you need to freeze something, should you fill the container to the top? Why or why not?

SCIENTIFIC METHODS SELF CHECK

✔ Did I **measure** exactly the same amount of water into each jar?

✔ Did I **predict** what would happen before I placed the jar inside the freezer?

✔ Did I **write** down my predictions and observations?

Erosion

Find Out
- What moves sediment
- How erosion changes Earth's surface
- What causes the most erosion

Vocabulary
erosion
deposition
glaciers
sediment

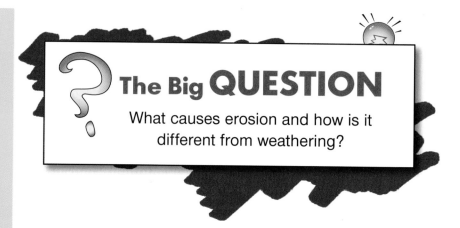

The Big QUESTION
What causes erosion and how is it different from weathering?

You know weathering breaks down rocks and causes changes in Earth's surface. Weathering is responsible for the potholes in the roads around you. Weathering can also wear down the stones and buildings in your area. But when a crack forms in the sidewalk, where do the sand and pebbles go? When weathering breaks the edges off sharp stones, what happens to the little broken bits and pieces? Erosion carries them away.

Moving Sediments

Weathering and erosion are natural processes that work together to change Earth's surface. First, weathering breaks down minerals and rocks into small pieces of sand, silt, or dust. Then, the process of **erosion** moves weathered particles away and drops them somewhere else. Forces of erosion are constantly moving and dropping loose, weathered materials. The natural forces that cause erosion are gravity, glaciers, wind, running water, and waves.

Agents of Erosion

Gravity

What happens if you drop a rock at the top of a very steep hill? The force of gravity will cause the rock to tumble down the hill. If you drop enough rocks, you will end up with a pile of rocks at the bottom of the hill. This buildup of moved particles is called **deposition.** Deposition can create new hills or even mountains if the particles are large enough. Gravity as a force of erosion is more obvious when there are steep hills or mountains. Weathering can break apart the underside of steep cliffs, and the remaining rocks will be pulled down by gravity.

Gravity will eventually erode this rock.

Glaciers

Another force that causes erosion is glaciers. **Glaciers** are large masses of slowly moving ice. They form when large amounts of snow are compressed into dense ice. Glaciers form in areas where snow does not melt as fast as it accumulates. Because of gravity, glaciers move slowly downhill, breaking and dragging rocks and other materials and depositing them somewhere else. You might expect that glaciers only form where the weather is constantly cold, like Antarctica, but they can also form in the very high mountains of warmer climates where snow piles up winter after winter.

Glaciers can move and drag rocks along, depositing them somewhere else.

Some glaciers, such as those in Antarctica, move very slowly, only about 2.5 cm a day. Others, like some in Alaska, may move 3 m a day. The speed at which a glacier moves is determined in part by how steep and rough the surface is under the glacier. The amount of water at the base of the glacier also affects the speed of a glacier. More water will speed the flow of the glacier downhill.

Wind

Wind can also move small particles broken down by weathering. When **sediment,** or loose materials, is small enough, the wind can pick it up and move it along. Unlike gravity or glaciers, wind cannot move large boulders. But when many small pieces of sediment are moved over many years, great changes in Earth's surface can occur. Wind erosion can cause great changes in desert areas. Rocks eroded by windblown sand can take on very distinct shapes. When the wind blows against one side of a rock, that side can become very worn and flat.

Desert winds can move huge amounts of sand as they shape and reshape the landscape of the desert. Sand dunes can move or migrate across the desert because of wind erosion. When the wind blows, it picks up sand and carries it over the top of the dune. As the wind blows down the other side of the dune, the sand is deposited. This constant process of picking up and depositing can cause the dune to migrate.

Sands dunes can move or migrate because of wind deposition.

Water—The Most Powerful Agent

Running Water

Of all causes of erosion, running water is the most powerful. Streams and rivers can have enough force to move small pieces of sand and even huge boulders. Running water can carve out canyons, valleys, and gullies. The Colorado River shows the extreme power of erosion by running water. As the river flows from higher levels to lower levels, the water picks up sediment and carries it downstream. Through

The running water of the Colorado River has helped to carve the Grand Canyon.

Which stream is carrying more sediment? Which stream is causing more erosion?

this process, the Colorado River has been a major force in carving out the landform we call the Grand Canyon.

The flow of the water in streams can influence the rate of erosion. Water that is flowing at higher speeds can carry more sediment. The more sediment the water carries, the faster erosion will take place. This is why steep mountain streams cause faster erosion than shallow, gentle streams. However, after severe storms, such streams may flow at a much greater speed. This increased flow will carry away more sediment. Because of the increased flow, the effects of erosion can be more evident after floods and storms.

Waves

Waves along a coast of an ocean or along a lakeshore can also cause erosion to take place. The slow and steady movement of waves carries along sand and bits of shells, moving them from one beach area to another. During clear weather, much of the sediment is picked up and replaced by other sediment. This wave motion only causes gradual erosion. During storms, however, large waves may pick up and carry away huge amounts of sediments, leaving behind eroded and rocky beaches. Severe storm waves have enough force to move large rocks and greatly change the shape of a coastline.

Erosion can be a slow and gradual process or it can be a dramatic and damaging process. Human activities can also speed up or slow down erosion. Can you think of any examples of erosion around you?

During a storm, waves can cause severe erosion.

CHECKPOINT

1. What moves sediment?
2. How does erosion change Earth's surface?
3. What causes the most erosion?

 What causes erosion and how is it different from weathering?

ACTIVITY
Eroding with Water

Find Out
Do this activity to see how water can cause erosion.

Process Skills
Constructing Models
Hypothesizing
Communicating
Observing

WHAT YOU NEED

plastic shoe box

water

graduated cylinder

clay

sand

Activity Journal

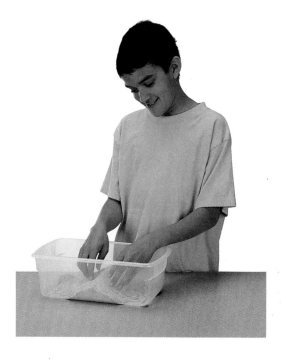

WHAT TO DO

1. In the shoe box, make a mound from a mixture of sand and clay.
2. Write a hypothesis about how water will affect your mound.

B78

3. Slowly pour 10 mL of water on top of the mound. **Observe** what happens to the sand and clay. **Record** your observations.

4. Compare your mound with how the mounds of the other groups look. Write any similarities.

CONCLUSIONS

1. What effect did the water have on the mound of soil?
2. How did the shape of the mound change?
3. What does the water in the cylinder represent on Earth?

ASKING NEW QUESTIONS

1. If you pour the water out quickly, what do you think will happen to the mound? Try this.
2. Where can you find examples of water causing erosion around your school?

SCIENTIFIC METHODS SELF CHECK

✔ Did I write a **hypothesis**?

✔ Did I **model** water erosion by pouring water onto the soil in the box?

✔ Did I **write** down my observations and compare our results with the other groups?

Catastrophic Events

Find Out
- What causes avalanches and landslides
- How earthquakes and volcanoes are alike

Vocabulary
avalanche
landslide
earthquake
aftershocks
volcano
lava

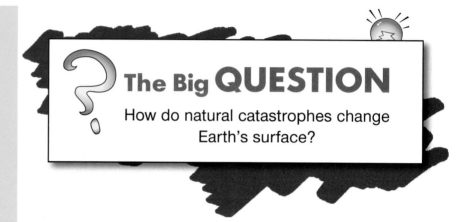

The Big QUESTION
How do natural catastrophes change Earth's surface?

Erosion can change the surface of Earth slowly. But other naturally occurring events can make big changes to Earth's crust, sometimes causing it to move. Often these changes are very destructive to whatever is living in the area, including humans. But after the destruction, these changes bring new features to the land and chances for new life-forms to move into the area.

Avalanches and Landslides

Weathering—caused by wind, water, and ice—and erosion—caused by gravity, running water, glaciers, wind, and waves—slowly change the surface of Earth. Gravity can cause erosion without any extraordinary changes in weather. Its force naturally pulls materials downhill, sometimes slowly, one rock or piece of rock at a time. But sometimes the movement is fast and violent when it takes the form of landslides or avalanches.

An **avalanche** is a sudden, downward movement of snow and ice. An avalanche can send tons of snow hurtling downhill at speeds of 160 kilometers an hour. In areas where avalanches are common, people sometimes create a "controlled" avalanche by setting explosives so they can control the slide. This reduces the buildup of snow in the mountains that might break loose and destroy the communities in the valley below.

Snow is not the only thing that can be pulled downward by gravity. A **landslide** is a sudden movement of soil, rock, or mud down a slope. Landslides happen when water gets into the cracks among rocks and soil on hills with steep slopes. Landslides sometimes follow rainstorms that have loosened the rocks on the hillsides. Water can also wear away the supporting structures under cliffs and hillsides, making it easier for them to collapse. Forest fires can destroy plant life on slopes, which makes the soil more likely to wash away and turn into a landslide.

A huge avalanche in Lauterbrunnen, Switzerland

Hurricane Mitch caused severe landslides in Nicaragua.

Earthquakes

An earthquake can occur suddenly, without warning. It can make what seems like solid ground roll like an ocean wave. Earthquakes have caused skyscrapers and bridges to collapse in seconds. Homes and entire cities can be destroyed by earthquakes.

An **earthquake** is a sudden movement of the rocks of Earth's crust. The movement is caused by a release of energy that has built up along a fault. Faults are areas on Earth's crust where the rocks have broken and shifted.

An earthquake can last several minutes and is often followed by **aftershocks,** which are smaller earthquakes radiating from the center of the first earthquake. Think about dropping a stone in water. The ripples in the water are like the aftershocks of an earthquake. They can occur for several days after the first earthquake is over.

Ripples in the water are like aftershocks following the first earthquake.

Earthquake damage in San Francisco in 1989

Some areas on Earth have more earthquakes than other areas. Earth's greatest earthquake belt is around the rim of the Pacific Ocean. Eighty percent of the world's earthquakes occur in this area. Along this belt lies the state of California, which has had many damaging earthquakes. The earthquake belt is a region of growing mountains and deep ocean trenches.

Measuring Earthquakes

Scientists measure earthquakes with a machine called a seismograph. The scale used for measuring earthquakes begins with a measurement below 3, meaning very small strength, and goes to 8, meaning great in strength. Earthquakes that measure at level 1 or 2 are so small they can hardly be noticed by people. These very small earthquakes happen every day. Strong earthquakes happen less than 150 times a year. They have a measurement of 6 or higher.

Earthquakes		
Descriptor	**Magnitude**	**Average Annually**
Great	8 and higher	1
Major	7–7.9	18
Strong	6–6.9	120
Moderate	5–5.9	800
Light	4–4.9	6200 (estimated)
Minor	3–3.9	49,000 (estimated)
Very Minor	1–3	about 1000 to 2000 per day

Source: U.S. Government Services, National Earthquake Information Center

Severe Earthquakes		
Date	Location	Richter Scale
1906	Coast of Colombia	8.9
1906	Jammu and Kashmir, India	8.6
1920	Kansu Province, China	8.5
1929	Fox Islands, Alaska	8.6
1933	North Honshu, Japan	8.9
1941	Coast of Portugal	8.4
1952	Kamchatka, Soviet Union	8.9
1964	Prince William Sound, Alaska	8.5
1985	Mexico	8.1
1991	Costa Rica	7.4
1993	Northern Japan	7.8
1994	Los Angeles, California	6.4
1995	Kobe, Japan	7.2
1999	Colombia	6.0

Predicting Earthquakes

Scientists use laser beams to measure movement within Earth. Sometimes scientists can use this information to predict when strong earthquakes will occur. Many lives could be saved if earthquakes could be predicted.

Scientists also study the behavior of animals to predict earthquakes. In 1974, in Haicheng, China, people noticed that snakes crawled out of their dens where they usually spend the winter, even though it meant they would freeze to death. Pigs tried to climb walls and refused to eat. Rats ran into crowded streets. Shortly after these animals acted so strangely, there were many small earthquakes. When the animals acted this way again two months later, many people left the city. They were lucky because a much more severe earthquake soon followed. Scientists are trying to figure out why some animals seem to be able to sense an earthquake before people can.

Volcanoes

Volcanoes and earthquakes often happen near one another. A **volcano** is an opening in Earth's crust through which steam, other gases, lava, and ashes come out. Volcanoes also occur most often in the area around the Pacific Ocean. Melted rock that is beneath Earth's crust rises to the surface through volcanoes. When a volcano erupts, the melted rock is joined by gas, steam, and ash. **Lava,** which is the melted rock when it is above ground, is sometimes very thick and flows slowly as it comes out of the volcano. At other times, it can be very thin and flow very quickly. Lava can erupt through cracks in the sides of the volcano as well as shoot out of the top.

Like earthquakes, volcanoes are a release of energy from Earth's interior. Volcanoes differ greatly in size, structure, and how often they erupt. Some volcanoes become mountains. Most volcanoes are under the ocean. The most forceful volcanoes can have an enormous impact on the life around them. The hot gases, lava, ash, and rocks that come out of the top of the volcano often destroy whatever is on or near it. However, as the lava cools, new rocks are formed. The processes of weathering and erosion will begin to break down these new rocks so that the surface of Earth will continue to change.

Lava flows and fountains on La Reunion Island

CHECKPOINT

1. What causes avalanches and landslides?
2. How are earthquakes and volcanoes alike?
 How do natural catastrophes change Earth's surface?

ACTIVITY
Making an Earthquake

Find Out
Do this activity to see how an earthquake can disturb buildings and other structures.

Process Skills
Constructing Models
Communicating
Observing

What You Need

medium baking pan

1.5 L water

28 toothpicks

gallon self-sealing plastic bag

16 pea-sized clay balls

Activity Journal

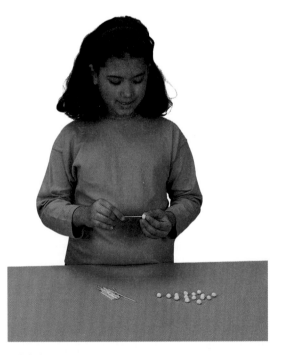

What to Do

1. Using the toothpicks and clay, **construct** a model of a building three stories tall. **Make a drawing** of your building.
2. Place the pan on a table. Pour the water into the self-sealing plastic bag and seal the bag.
3. Put the bag in the pan. Place your building on the bag of water.

4. Then create an earthquake. Start with a minor earthquake and work your way up to a strong earthquake. Move the pan a little bit. Then move it a little more. Then really shake it.

5. Record what happens to your building as you increase your earthquake's intensity.

CONCLUSIONS

1. How did the building change as you moved the pan at different speeds?
2. What happened to your building after you stopped moving the pan?

ASKING NEW QUESTIONS

1. How could you design your building differently to decrease the amount of damage during the simulated earthquake?
2. Do you think you would feel an earthquake more on the top floor of a building or in the basement?

SCIENTIFIC METHODS SELF CHECK

✔ Did I **construct** a model of a building?

✔ Did I **make a drawing** of my model?

✔ Did I **observe** the building when I moved the pan?

✔ Did I **write** down my observations?

Review

Reviewing Vocabulary and Concepts

Write the letter that completes each sentence.

1. When substances that make up a rock are changed, it is called ___.
   - **a.** landslide
   - **b.** physical weathering
   - **c.** chemical weathering
   - **d.** aftershock

2. When a frozen substance changes to a liquid by warming, this is called ___.
   - **a.** thawing
   - **b.** freezing
   - **c.** physical weathering
   - **d.** chemical weathering

3. ___ moves weathered particles and drops them somewhere else.
   - **a.** An aftershock
   - **b.** Chemical weathering
   - **c.** Sediment
   - **d.** Erosion

4. A buildup of particles that have moved from one place to another is called ___.
   - **a.** deposition
   - **b.** aftershock
   - **c.** thawing
   - **d.** erosion

5. Large masses of slowly moving ice are called ___.
   - **a.** avalanches
   - **b.** glaciers
   - **c.** volcanoes
   - **d.** aftershocks

6. ___ are snow and ice moving downward at great speeds.
   - **a.** Landslides
   - **b.** Earthquakes
   - **c.** Volcanoes
   - **d.** Avalanches

7. A quick movement of soil or rock down a slope is called ___.
   - **a.** an earthquake
   - **b.** thawing
   - **c.** sediment
   - **d.** a landslide

8. An opening in Earth's crust through which steam, lava, and ashes come out is called ___.
   - **a.** a glacier
   - **b.** sediment
   - **c.** weathering
   - **d.** a volcano

Understanding What You Learned

1. Describe the differences between a rock changed by physical weathering and a rock changed by chemical weathering.
2. Describe how hills and mountains change because of weathering and erosion.
3. What is the difference between an avalanche and a glacier?
4. How can earthquakes be predicted?

Applying What You Learned

1. How does water weather the rocks on mountains very quickly?
2. What evidence of weathering and erosion might you find in a desert area?
3. How can different forces cause changes in Earth's surface?

For Your Portfolio

Make a flip book showing the changing motion of an exploding volcano, earthquake, or avalanche. Use 10–20 sheets of paper and draw pictures so that the images align from one page to the next. Color your pictures and show evidence of Earth's changing surface. Flipping the pages should create a moving picture.

CHAPTER 4
ROCKS, MINERALS, AND SOILS

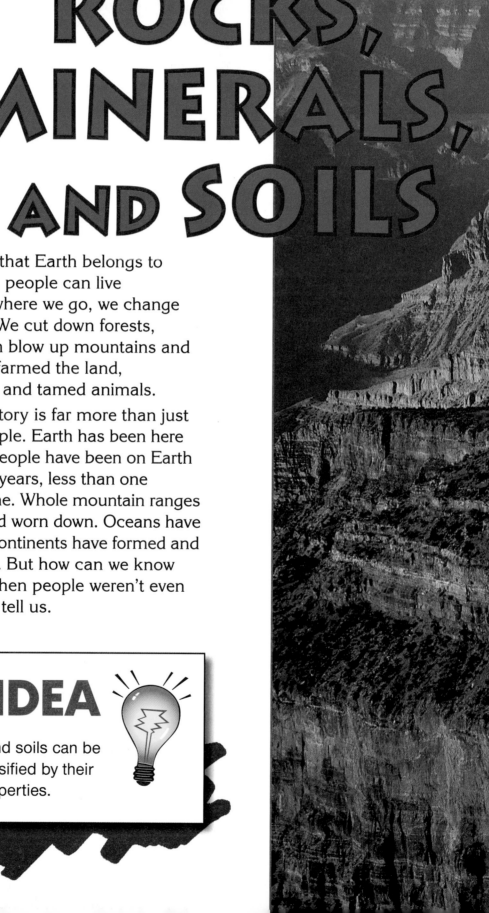

It's easy to think that Earth belongs to humans. After all, people can live anywhere. Everywhere we go, we change things to suit us. We cut down forests, create lakes—even blow up mountains and islands. We have farmed the land, developed plants, and tamed animals.

But Earth's history is far more than just the history of people. Earth has been here 4.6 billion years. People have been on Earth only a few million years, less than one percent of that time. Whole mountain ranges have grown up and worn down. Oceans have risen and fallen. Continents have formed and then broken apart. But how can we know what happened when people weren't even alive then? Rocks tell us.

The Big IDEA

Rocks, minerals, and soils can be identified and classified by their physical properties.

CHAPTER SCIENCE INVESTIGATION

Make a compost pile so that you can understand how soil is formed. Find out how in your *Activity Journal*.

Lesson 1

Rock Types and Formation

Find Out
- How the three rock types are formed
- What the rock cycle is
- How fossils help to tell the age of rock layers

Vocabulary

rock cycle
relative dating
fossils
index fossils
geologic time table

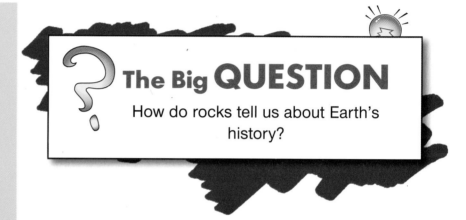

The Big QUESTION

How do rocks tell us about Earth's history?

Have you ever gone into the kitchen and mixed lots of different ingredients? You might have made a casserole, a cake—or a mess! Nature does the same thing. But when nature mixes ingredients, it sometimes results in a variety of rocks being formed.

Different Types of Rocks

Rocks can be grouped into three types: igneous, sedimentary, and metamorphic. Igneous (ig′ nē əs) rocks are formed by cooling liquid rock called magma. When magma bursts through cracks in Earth's crust, it becomes lava. As lava cools, it hardens to form igneous rocks. Sometimes magma cools slowly underground, and there is time for large crystals to develop in the igneous rock. Granite, a coarse-grained rock, is formed this way.

B92

When lava cools quickly in the air on Earth's surface, there is only time for small crystals to form. We get basalt, a fine-grained rock, this way. Obsidian, a natural glass, is formed when lava cools so quickly there is no time for any crystals to form.

Sedimentary (sed′ ə men′ tə rē) rocks form when loose materials, such as small rocks or pieces of dead plants and animals, are pressed together. Layers must build up over millions of years, with the upper layers pressing down on the lower layers. If the sediments are very small, such as particles of clay, pressure alone can form rocks.

Larger sediments, such as gravel, must be cemented together. This can happen when water containing minerals flows through open spaces in the sediments. The minerals are left in these open spaces and around the sediments. The sediments then stick together like glue to form solid rock.

Metamorphic (met ə mor′ fik) rocks are formed when igneous or sedimentary rocks are changed by great heat or pressure—or both. A rock can change when it is exposed to intense heat. For example, the sedimentary rock sandstone becomes quartzite when it is exposed to intense heat. Limestone, another sedimentary rock, becomes marble. When shale undergoes great pressure, such as that found inside Earth's crust, it changes into slate.

Granite

Obsidian

Limestone

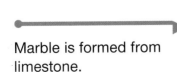

Marble is formed from limestone.

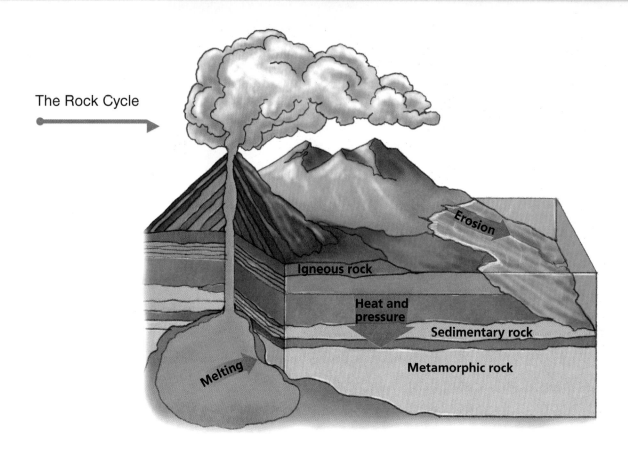

The Rock Cycle

The Rock Cycle

Although it takes millions of years, rocks are always in the process of changing. They form and wear down and then form again. This process is called the **rock cycle.**

The rock cycle starts when volcanoes make igneous rocks at Earth's surface. Then the forces of weathering break the igneous rocks down into bits and pieces. These pieces are eroded and deposited to form sedimentary rocks. These rocks become buried under many layers of other rock. When buried deeply, they undergo intense heat, pressure, or both to become metamorphic rocks. Then, they may melt as they are pushed further into Earth's crust and start over again as igneous rocks.

A Record in Rock

If you have ever driven on a road cut through a mountain, you have seen rock put down in layers. Think of these rock layers as being like a stack of newspapers you are collecting to recycle. The oldest newspaper is at the bottom of the pile. Each day you add a paper, with the newest one always being on top. In the same way, the older layers of rock in a formation are always underneath the younger layers because they were put down first.

You can't lay down a new layer of rock between or under two that are already in place without disturbing the original layers. Those layers can only be disturbed by large movements, such as earthquakes or landslides, which break a formation apart. Then the layers might be mixed up when the rocks re-form. Imagine that someone knocked over your stack of newspapers. Some of the dates might get out of order even if you try to push them back into one pile.

Figuring out the order in which events occurred is called **relative dating.** We can't tell how old rock is just by looking at the layers. But we can tell which layer is older than another by the order they are in. This type of dating helps scientists to understand the order of events in Earth's history.

New layers of rock form on top of older layers of rock.

Fossils and Earth's History

Scientists can get other clues to Earth's history from the fossils they find in the rock layers. **Fossils** are traces or remains of plants or animals. They are formed when parts of plants or animals are preserved in mud. When the mud gets covered, it eventually turns to rock. Fossils can tell scientists which plants and animals were alive when that rock layer was formed. They can also tell what the world was like at that time. For example, suppose you found fossil fern leaves in a rock layer. We know the kind of climate and habitat that ferns live in today, so ancient ferns may have lived in the same kind of environment.

Index fossils are fossils that can be used to help scientists determine the age of rock layers. They are the remains of animals or plants that lived only during a short time period. For index fossils to be useful, they have to be easy to identify. They also need to be found in different types of sedimentary rock over wide areas. A good example is the trilobite (trī′ lə bīt′).

Suppose you found a trilobite fossil in a layer of shale, and above it you found a layer of sandstone that did not have a fossil. What could you say about the age of the sandstone? You might guess that the sandstone was younger. And if you found a trilobite in another rock formation hundreds of miles away, what could you say about it? You could say that rock was about the same age as the shale.

A trilobite fossil can help date the age of a rock layer.

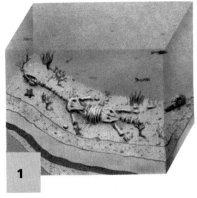
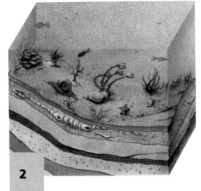

→ The fossilization of a Plesiosaur, a reptile from the Mesozoic Era

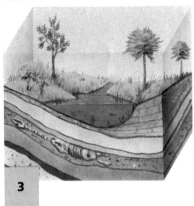
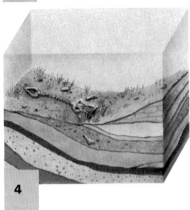

Index fossils have helped scientists develop a **geologic timetable.** The geologic timetable identifies four different times in Earth's history: Precambrian, Paleozoic, Mesozoic, and Cenozoic. Many plants and animals became extinct at the end of each time period. By studying the fossils found in rock layers from these times, scientists can tell which animals and plants were alive during a particular time.

CHECKPOINT

1. Identify the three basic kinds of rocks and tell how they are formed.
2. Describe the rock cycle.
3. How can fossils help to tell the age of rock layers?
    How do rocks tell us about Earth's history?

B97

ACTIVITY

Classifying Rocks

Find Out
Do this activity to learn how to classify rocks using texture and other properties.

Process Skills
Observing
Communicating
Classifying
Inferring

What You Need

rock samples: granite, obsidian, pumice, basalt, slate, marble, gneiss, schist, shale, limestone, and sandstone

hand lens

Activity Journal

What to Do

1. Carefully observe the whole group of rocks. Don't use the hand lens yet. Sketch and label each rock.

2. Write the general texture of each rock. Is it smooth or rough? Is the rock layered? Are there any holes in the rock?

3. Then group the rocks according to their similar properties.

4. **Observe** each rock with the hand lens. Add to your sketches, if necessary.
5. After using the hand lens, would you change your original groupings?

Conclusions

1. Did you change your original groupings? Why or why not?
2. Are your groups like your classmates' groups? Is there only one way to group the rocks? Explain.
3. Did some rocks fall into more than one category? Why or why not?

Asking New Questions

1. Explain why you grouped your rocks the way you did. What characteristics did you look for?
2. What other characteristics could you look for?
3. After looking at the different properties of rocks, try to describe how each rock could have been formed.

> ### SCIENTIFIC METHODS SELF CHECK
> ✔ Did I **observe** the rock with and without a hand lens?
> ✔ Did I **record** the texture of each rock?
> ✔ Did I **classify** rocks by their properties?
> ✔ Did I try to **explain** how the rocks were formed by looking at their observable properties?

Lesson 2

Earth's Minerals

Find Out
- What the properties of minerals are
- How to identify minerals
- What rocks and minerals are used for

Vocabulary
minerals
crystals
luster
streak

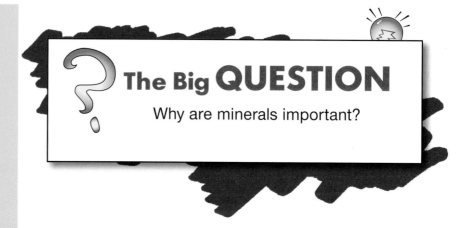

The Big QUESTION

Why are minerals important?

You've learned how rocks are formed, but have you thought about what's inside a rock? What makes some rocks colorful? What makes some rocks hard and others soft? What makes a few rocks more valuable than all the others? In this lesson, you will learn about the minerals that make up rocks and how they are useful to us.

Properties of Minerals

Rocks are made of one or more minerals. **Minerals** are solids that form naturally and have a specific chemical makeup and physical form. Minerals are made of one or more of Earth's elements, such as silicon, oxygen, carbon, and iron. Elements combine to form about 500 different minerals, although only about 100 of these are common. Minerals can be found in caves, streams, rivers, and oceans. Many occur in Earth's crust close to the surface. Some common minerals are quartz, talc, salt, and graphite.

A substance must have five characteristics to be considered a mineral. First, it must be a solid, not a liquid or a gas. Second, it must be made from nonliving things. Third, it must occur naturally on Earth. Steel is not a mineral because it is made by humans. Fourth, it must be a single substance that is always made from the exact same elements in the same proportions. Sand is not a mineral because samples from different places usually contain different elements. Fifth, a mineral's atoms must be arranged in repeating patterns called **crystals.** These crystal structures are the building blocks from which minerals are made.

Each mineral is different from the others, because it is made of different materials. Minerals might differ in color, hardness, and many other ways. These differences or properties help scientists identify them.

Crystals of table salt are cubic.

This granite rock contains many different minerals such as feldspar, mica, and quartz.

B101

Identifying Minerals

Mohs Scale of Mineral Hardness

Hardness	Mineral	General Features
1	Talc	soft, flaky, slightly greasy
2	Gypsum	can be scratched by your fingernail
3	Calcite	can be scratched by a penny
4	Fluorite	can be scratched by a knife
5	Apatite	barely scratched by a knife
6	Orthoclase	can scratch glass easily
7	Quartz	can scratch a steel file
8	Topaz	can scratch a quartz
9	Corundum	can scratch a topaz
10	Diamond	cannot be scratched

Minerals can be tested for hardness, luster, color, and the way they break. The hardness of a mineral is important to know if you plan to use it in manufacturing. The hardest mineral of all is diamond. It is made from one single element, carbon. It is the hardest natural substance because of the way the carbon atoms are combined in it. You might think that diamonds are just used to make beautiful jewelry, but because of their hardness, they are also used for many cutting tools. Drill tips can be made with diamonds to allow them to cut through other substances.

In 1812 a scientist named Friedrich Mohs developed a scale to rate the hardness of minerals in comparison to each other. On his scale, talc—one of the softest minerals—is rated 1. It can easily be scratched by your fingernail. The hardest, of course, is diamond. It is rated 10. Mohs used eight other common minerals with different hardnesses for the numbers in between. Each mineral can scratch minerals that have a lower number. All minerals can be placed somewhere on this scale by comparing them to see which minerals they can scratch. For example, an emerald would rank about 7.5 because it can scratch quartz (7), but not topaz (8).

Another property minerals are tested for is how they reflect light, which is called **luster**. A mineral's luster can be described as metallic (bright and shiny like a metal) or nonmetallic (dull or pearly). Pyrite is a mineral that is commonly called "fool's gold" because its metallic luster looks like the shiny luster of real gold. In contrast, talc looks pearly, and quartz looks glassy.

Color is also used to identify minerals, though it is not always reliable. Air and water can change the appearance of some minerals. However, if you rub a mineral across a white, unglazed ceramic tile, you will see its true, or streak, color. The **streak** of the mineral always stays the same.

You may have noticed that minerals have different shapes. This is because minerals break in different ways. Some minerals, such as mica, break in one direction, leaving a flat side. Other minerals break into different shapes. Calcite can break in three directions, splitting it into blocks. Mineral gemstones, such as diamonds and emeralds, are broken in special ways to show off their beauty.

Talc

Pyrite

Black quartz

Diamonds are cut to make them beautiful for jewelry.

Mica breaks into sheets.

Calcite breaks into blocks.

B103

Using Rocks and Minerals

Arrowheads were made of flint.

Gypsum can be found in the walls of many buildings.

Rocks are all around you. Since early times, people have found many uses for rocks and the minerals contained within them.

Many thousands of years ago, people made weapons and tools from rocks. A spear would have a point chipped from a rock such as flint. A hard, tough rock, such as granite, might be used for a hammer. In ancient Rome, workers cut blocks of rock to make roads. Some of these roads can still be used. Today, rocks are used in bridges, statues, roads, and buildings. They are also widely used for decoration.

The properties of minerals help determine how they are used. You have already learned that the hardness of diamonds makes them useful for drilling. Soft minerals are useful, too. The black stuff in your pencil is the mineral graphite. Because it's soft, it will rub off on your paper as you write. The line you draw on your paper is the streak color of graphite.

Quartz and garnet are used in sandpaper, and gypsum may be used in the plaster on the walls around you. Many watches contain quartz crystals because they vibrate at a regular rate when electricity is passed through them. Even the salt you sprinkle on your food is a mineral.

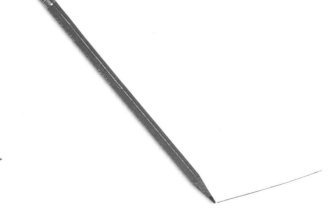

Graphite is a mineral.

Minerals such as iron, copper, and aluminum are used to make everything from pots and pans to cars and machinery. They are also used to make fabrics and medicines. The next time you brush your teeth or wash your hair, read the toothpaste or shampoo container to see if you recognize the names of any minerals there.

Early people used minerals for paint. Many other civilizations have also used minerals for glazes and coloring. We can still see some of these paintings today, and they can help us learn about the lives of ancient peoples.

Sometimes, the minerals we use can be harmful to our health. Asbestos is a grayish-white mineral that in some forms can cause a lot of health problems. Because it doesn't burn, it was used in ceilings for fire protection. Scientists now know that when asbestos comes loose and the fibers float in the air, it can damage our lungs and cause health problems. Because of this, it is being carefully removed from many buildings.

Native American painters used minerals for paint.

Asbestos is a mineral that can be harmful to your health.

CHECKPOINT

1. What is a mineral?
2. How can you identify minerals?
3. How do we use minerals today?
 Why are minerals important?

ACTIVITY

Naming Minerals

Find Out
Do this activity to identify minerals by studying their properties.

Process Skills
Communicating
Observing
Defining Operationally
Classifying

What You Need

mineral samples: quartz, calcite, gypsum, pyrite, talc, halite, specular hematite, galena, fluorite, orthoclase or other feldspar, a form of mica, hornblende

steel file
penny
glass plate
unglazed, porcelain tile
Activity Journal

What to Do

1. List each mineral and the following headings: "Luster," "Streak Color," "Heaviness," and "Softness/Hardness Rank."

2. **Observe** each rock. **Record** the luster of each mineral using words like *dull, shiny, glassy, milky,* or *pearly.*

3. Rub each mineral across the tile. **Record** the color of the mark on the tile, if any.

4. **Decide** how heavy your samples are by tossing them gently from hand to hand. One or two minerals should stand out as being much heavier or lighter than the others. **Record** your observations.

5. **Rank** your samples from softest to hardest.

   a. Start by trying to scratch each mineral with your fingernail. Set aside those minerals that you're able to scratch. These are considered soft minerals.

   b. Next, see if you can scratch the remaining minerals with a penny. Set aside those that you can scratch.

   c. Next, see if a steel file will scratch the remaining minerals. Set aside those that cannot be scratched. Are these minerals softer or harder than the file?

   d. Finally, see if the minerals you have left will scratch a glass plate. Set aside those that do not.

   e. Within each group, **test** to see if one mineral scratches another. **Rank** and **record** the minerals in a group from softest to hardest. **Compare** your group's ranking with the other groups. **Discuss** any differences.

CONCLUSIONS

1. Did any minerals produce a surprising streak?
2. Which samples stood out as being heavy? Can you suggest a reason why this might be?
3. Which sample was the hardest? Which was the softest?
4. Which samples could be scratched by a fingernail? By a penny? By the steel file? Which could scratch glass?

ASKING NEW QUESTIONS

1. Which tests seemed to be best for classifying minerals? Which tests were not so good?
2. Why are some of these minerals harder than others?

SCIENTIFIC METHODS SELF CHECK

✔ Did I **observe** and **record** the testing of minerals?

✔ Did I **classify** minerals according to their properties?

B107

Soils

Find Out
- What soil is made of
- How soil forms
- How the properties of soil can be compared

Vocabulary
soil
humus
sand
silt
clay
water retention

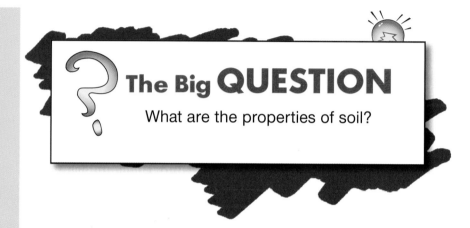

The Big QUESTION

What are the properties of soil?

When you think of soil, you may think of dry, red clay or damp, black potting soil sold in bags at the local store. The color, the feel, and even the smell of these two soils are very different. But they are both soil.

What Is Soil?

Soil is a mixture of rock, plant material, animal material, water, and air. To make soil, you must begin with rocks.

Soil starts as rocks on Earth's surface. First, weathering and erosion work on them. Ice, rain, and wind beat upon the rocks. Water can seep into cracks and then freeze, breaking the rocks apart. Acid in the rain can dissolve minerals in the rocks. When rocks warm during the day and cool at night, they may begin to crack and break. Gradually, all of these forces of nature break the rocks down into tiny particles.

Various plants are able to grow in the cracks of the rocks. Also, mosses and lichens are able to grow on rocks. These living organisms help break the rocks apart. When they die, the plant material mixes with the tiny rock particles. Animals that live near the rocks will eventually die, and their body parts will mix with the tiny rock particles.

Soil Formation

Soil formation depends on five main factors. The first is time. It takes lots of time to make soil. Think about how long it would take to break even one large boulder into tiny particles by just having the wind blow sand against it or rain beat down on it. It can take up to 1000 years for just 1 cm of soil to form.

The type of rock in an area also affects soil formation. The harder the rock is, the more time it will take to break down.

Climate, too, affects soil formation. In a cool, dry climate, erosion happens more slowly simply because there is less rain and wind. In a hot, moist climate, there is more water, so the formation of soil can happen more rapidly.

The slope of the land has a great effect on the formation of soil. If land has a flat surface, a deep layer of soil can form because running water may have a harder time eroding it. The soil-making materials are not carried away by water. Sloping land allows water to flow quickly, carrying soil and rock particles away.

Finally, plants and animals affect soil formation. Decomposed plant and animal matter, called **humus** (hyōō′ məs), mixes with tiny rock particles to form soil. If there are

Humans are dependent on soil.

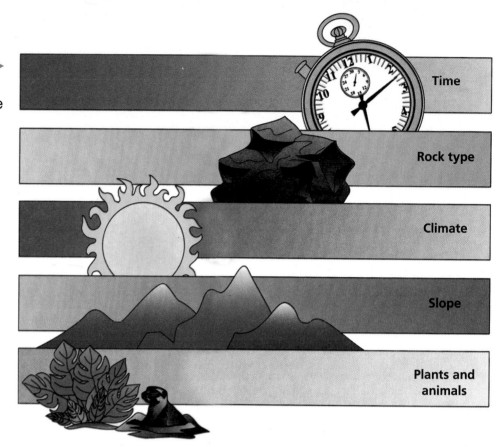

Soil formation depends upon five factors.

many animals that live in the soil, such as earthworms, soil forms more quickly. When animals tunnel through the soil, they bring air and moisture beneath the surface of land, which speeds soil formation.

Different Kinds of Soil

Did you know that there are thousands of kinds of soil on Earth? Soils can be black, red, brown, gray, yellow, or a combination of these colors. Soils can also have a variety of textures from moist and sticky to gritty. The type of soil formed depends on the type of rock it came from. It also depends on the kinds of plants and animals that contributed to its formation.

The different soils can all be divided into three types: sand, silt, or clay. There are advantages and disadvantages of each type of soil.

Sand has large grains of rock big enough for you to see. It is also the heaviest of the three soil types. Its texture is gritty, and the grains of sand usually do not stick to each other. Water can pass through sandy soil easily and quickly. This means that it can be hard to grow plants in sandy soil. Not enough moisture will stay in the soil.

The grains in **silt** can only be seen through a magnifying glass or under a microscope. Its texture is powdery. It feels slippery but not sticky. It can retain or hold water better than sand, but since it is really fine, it erodes easily.

Clay has the smallest grains. They can only be seen under an electron microscope. When clay is wet, its grains stick to each other, and it can be molded with your hands. When it dries, however, it forms a hard brick. It is difficult for plants to push their roots through clay.

Humus helps to hold water and air in soil.

The best soil for growing things is a mixture of these three types. Sand lets air and moisture into the soil. This allows plants to get the nourishment they need and lets roots grow. Silt provides the bulk of the soil. And clay holds the soil together and keeps the water from escaping too quickly.

There is an easy way to get an idea of how much sand, silt, and clay a sample of soil has. Take a clump of soil about the size of a marble and get it wet. Then try to flatten it, like a ribbon. If it crumbles and won't form into a ribbon, there is probably a lot of sand in the soil. If you can form a ribbon, but it won't stay together, there is probably a lot of silt in the soil. But if it sticks together well, there is probably a lot of clay in it.

Water Retention and Chemicals

Knowing how much sand, silt, and clay is in soil is not enough for predicting how well plants will grow in it. Two other important properties of soil are water retention and chemical makeup.

Water retention describes how well a soil holds water. Plants can't grow without enough water. But it is also a problem if they get too much water. Scientists test to see how fast water soaks into soil, how much water the soil can hold, how long the water stays in the soil, and how quickly it passes through.

The next time it rains near you, watch to see if the water puddles up on the ground or soaks in. After it soaks in, how long does it take before the ground feels dry again?

Puddles are a sign of too much water in the soil.

As you have already learned, soil is made partly from rocks, and rocks contain minerals. The chemicals in the minerals can also affect how plants grow.

Many chemicals are good for plants. They provide food or nutrients for the plants. Nitrogen, phosphorus, potassium, and calcium are some of the important elements in soil that plants need. Other chemicals can hurt plant growth. Salt, for example, can hurt or kill many plants.

The chemical makeup of different soils can determine how well different plants grow in that soil. Strawberries, for example, grow well in acidic soil. But minerals can make the soil too acidic, causing plants to die. The right balance of chemicals and decomposed matter is needed for soil to support plant growth.

Strawberries grow well in acidic soil.

CHECKPOINT

1. What is soil made of?
2. What five factors influence soil formation?
3. How can different types of soil contribute to plant growth?
 What are the properties of soil?

ACTIVITY

Testing Soil Characteristics

Find Out
Do this activity to test soil characteristics such as texture, color, water retention, and hardness.

Process Skills
Observing
Communicating
Measuring
Controlling Variables
Interpreting Data

What You Need

newspaper, hand lens, masking tape, pencil, water, beaker, watch or clock with second hand, two large coffee cans, open at both ends, trowel, hammer, small wooden board, metric ruler, Activity Journal

What to Do

1. **Find** a place outside where plants don't grow very well. Find another spot where the grass is thick and green. Use the trowel to remove a section from each spot. Put a sample of soil from each spot on a newspaper and **examine** it with the hand lens. Try to poke your pencil into the soil at each spot. Which soil seems harder? **Record** your observations.

2. Using tape, mark the outside of each coffee can 5 cm from one end. Place a can at each location. Put the board on top of each can

and have one person hold the board while another person hammers the wood to push the can into the soil to the 5-cm mark.

Safety! *Tape both ends of the coffee cans. Use the hammer carefully. Keep your hands and fingers out of the way.*

3. Pour one beaker of water into each can, and time how long it takes the water to soak into the ground. **Record** the soaking time from each location.

CONCLUSIONS

1. What similarities and differences did you find between the two soil samples?
2. Where was water absorbed more quickly?
3. Remember the results when you poked the soil at each spot with your pencil. Do you think there is a relationship between soil characteristics and water absorption? Explain.

ASKING NEW QUESTIONS

1. Explain the relationship between your findings and the growth (or lack of growth) of plants at the soil locations.
2. How might you get grass to grow better on bare spots where little grass grows?

SCIENTIFIC METHODS SELF CHECK

✔ Did I **record** my observations?

✔ Did I **measure** the cans and the water?

✔ Did I **describe** the similarities and differences between the samples?

Soil as a Natural Resource

Find Out
- Why soil is important for plant and animal life
- How soil can be destroyed
- How soil can be conserved

Vocabulary
horizons
inexhaustible resources
soil conservation
composting

The Big QUESTION
Why is it important to preserve soil?

Imagine a world without soil. There would be bare rock and water. But there couldn't be much else! Without soil, plants could not grow. And without plants, no animals could live. Soil is important to all living things, so people need to work to preserve the soil that we have.

Soil and Plant Life

Think about all the things we get from soil. In addition to food, we get cotton and flax, which are used for clothing. Trees give us material for paper and wood to build houses. Many medicines come from plants. Almost everything in our lives can be traced back to plants. Hamburgers come from cattle and cattle eat plants. Wool for sweaters comes from sheep that have eaten leaves. People couldn't live without materials from plants grown in soil.

Layers of Soil

If you cut into Earth and remove a slice several meters deep, what would you find? Chances are you'd find several layers of soil. Scientists call these layers of soil **horizons.**

The top layer is the one that is most important to people and to plant growth. It is called topsoil, or the A horizon. It is usually dark in color and crumbly. It may contain bits of grass, small twigs, or plant roots. You might also find insects, earthworms, and seeds. This is the soil that most plants can grow in.

A layer of topsoil can go as deep as 5 m. The layer is usually somewhere between 40 and 55 cm. Topsoil differs from place to place on Earth. The kind of plants grown and the kind of animals that live in or nearby affect the soil. Even the types of rocks in the area can affect the soil.

Below the topsoil is subsoil, or the B horizon. Subsoil is usually finer and firmer than topsoil, though it may have grains of hard quartz in it. It is often mostly clay. Roots can grow down into subsoil, but there will not be any decomposing plant or animal material. It is often yellowish and is usually about 1 m deep.

The C horizon comes next. The top part is made of broken or chipped pieces of rock—weathered rock. This is the rock from which most of the soil will eventually come. If you go deep enough in the C horizon, you reach solid bedrock.

All types of soil eventually form layers called soil horizons.

Destroying the Soil

Good topsoil is needed to grow plants, but Earth is losing topsoil much faster than it can make it. Plants need soil, but soil also needs plants. Plants hold soil together, add decomposed material to the soil, and protect it from the forces of erosion.

Much topsoil is lost to erosion. The same forces that break down rocks, helping to make soil, can also destroy the topsoil. Wind can blow topsoil away. Rain and ice can loosen topsoil and then wash it away.

Where does the topsoil go? It depends on where it starts. Soil from the Rocky Mountains can be blown or washed into streams flowing down the mountainsides. These streams eventually flow into larger rivers, such as the Colorado River. Finally, the soil is deposited into the ocean.

Running water can erode topsoil much faster than it can be made.

Sometimes this process can help a region. Ancient Egyptian farmers looked forward to the Nile River flooding each year. The floods left rich new soil on their land. But many times, the soil is washed away to a place that can't be used for farming, such as the ocean. Sometimes it is left in a place where it causes trouble. Rivers can fill with this soil, becoming too shallow for boats to pass or for animals to live there.

There is always new soil being made, but it can take a thousand years to make 1 cm of soil. Erosion can destroy that same amount of soil in less than five years.

Soil is also lost because of the actions of people. Overfarming the land causes much of that loss. Farmers sometimes plant the same crops year after year in the same place. This will remove many useful minerals from the soil. Gradually, it becomes impossible to continue farming that land.

Livestock can destroy soil, too. If animals are allowed to, they will overgraze the plants in an area. Without plants, the soil is easily blown or washed away.

Logging and mining can also destroy topsoil. Loggers cut down trees to make room for farmland or to harvest trees for paper and lumber. Without trees, the soil is open to erosion. Even when new trees are planted, soil can still be lost. Tiny new trees can't prevent erosion as well as large trees can.

Strip mining can also leave the land open to erosion. Once the plants and trees are gone, nothing holds the soil together and protects it. Logging and mining equipment can also scrape away soil just by moving over it.

People also bury topsoil when they build roads and buildings. Many roads and parking lots used to be fields full of plants and trees.

The slash-and-burn method of clearing land for farming can erode valuable topsoil.

Saving the Soil

Soil is not an inexhaustible (in′ eg zä′ stə bəl) resource. **Inexhaustible resources** are things we use that can never run out. Sunlight is an inexhaustible resource because we can use it for many, many years to come. But we can run out of soil. In fact, some scientists warn that we are rapidly losing much of the world's farmland because of topsoil loss.

We need to conserve the soil we have. **Soil conservation** means to preserve and protect the soil and to use it wisely. People are working on many ways to keep soil from being destroyed.

Farmers and scientists have learned that rotating the crops that are grown in an area can help save soil. Different plants use different minerals in the soil. For example, a farmer could plant wheat and barley in a field one year. The next year the farmer could plant beans and peas. This rotation helps to return different minerals and nutrients to the soil.

Farmers can also help save the soil by adding nutrients and minerals to their fields. Farmers can plow some plants and roots back into the ground. The decomposing plant matter strengthens the soil.

People can contribute to soil conservation by planting trees and other plants. Trees are helpful in preventing soil erosion. Their roots not only hold the soil in place, but they also provide windbreaks, preventing soil from being blown away.

Crops can be planted across the hills, instead of up and down. This slows soil erosion by running water.

B120

People can also enrich poor soil by composting. **Composting** is a process for making the decomposing plant matter necessary for soil. To compost, people collect grass clippings, shredded branches, leaves, sawdust, and some food scraps in a large pile outdoors. They add water for moisture and turn the compost to mix in air. As the material decomposes, it makes a rich fertilizer for the land.

You can help, too. Why not plant a special tree this year on Earth Day or Arbor Day? Try growing flowers or vegetables in a garden. Don't throw away food. Learn how to recycle some of that food in a compost pile. Write a letter to your representative in Congress asking him or her to vote for bills that promote soil conservation. And always recycle as much as you can.

Humans share Earth with all living organisms. We are all bound together by Earth and by what happens to it. So let's use Earth's soil resources wisely!

This bucket has holes in the bottom. It will release water into the soil around the tree slowly. This method of watering the tree will result in less water being wasted and less soil erosion.

CHECKPOINT

1. Why would life be very different without soil?
2. What are some ways soil is destroyed?
3. What are some ways to conserve our soil?
 Why is it important to preserve soil?

ACTIVITY

Modeling Earth's Limited Resources

Find Out
Do this activity to find out how much of Earth's land is available for growing crops.

Process Skills
Constructing Models
Using Numbers
Observing
Communicating

WHAT YOU NEED

red, blue, yellow, and green modeling clay

a plastic knife

Activity Journal

WHAT TO DO

1. Mold the red clay into the shape of an apple. The red clay represents Earth.

2. Cut the red ball in half. Take one of the halves and cut it in half again so you now have two ¼ wedges. Stick one of the ¼ wedges back on the red ball.

3. Cover the outside of the ¾ red ball with blue clay. This represents the water covering Earth's surface.

B122

4. Cover the outside surface of the other ¼ wedge with yellow clay. This **represents** the land areas on Earth.
5. Now cut the ¼ land wedge into two ⅛ wedges.
6. Cut one ⅛ wedge into four equal sections to make four 1/32 wedges.
7. Cover the outside surface of one of the 1/32 wedges with green clay. This green section **represents** the land on Earth where all farming is done.
8. **Observe** the model that you have cut apart and **record** why you think there are more yellow sections than green sections.
9. Make a **graph** that shows the size of the different sections.

Conclusions

1. Why are there more yellow sections than green sections?
2. Why should we try to conserve topsoil for growing crops?
3. On which section of your model do you live?

Asking New Questions

1. Where else on Earth could people look to produce food?
2. What can we do to keep the green part of Earth from getting even smaller?

> **SCIENTIFIC METHODS SELF CHECK**
> ✔ Did I **construct a model** of the land and water on Earth?
> ✔ Did I **divide** the sections of the model correctly?
> ✔ Did I **observe** the different sizes of the sections?
> ✔ Did I **write down** my observations?

Review

Reviewing Vocabulary and Concepts

Write the letter of the answer that completes each sentence.

1. The process of rock changing from one form to another is the ___.
   - a. metamorphic cycle
   - b. sedimentary cycle
   - c. rock cycle
   - d. igneous cycle

2. Scientists try to figure out the order of events millions of years ago by ___.
   - a. measuring rock
   - b. weighing rock
   - c. striking on tile
   - d. relative dating

3. How quickly or slowly an igneous rock cools can influence the formation of ___.
   - a. sediments
   - b. crystals
   - c. layers
   - d. fossils

4. Solids that are formed of a single substance and have a specific chemical makeup are called ___.
   - a. minerals
   - b. rocks
   - c. dated
   - d. streaks

5. We call how bright a mineral shines its ___.
   - a. texture
   - b. hardness
   - c. luster
   - d. value

6. The true color of a mineral is its ___.
   - a. streak
   - b. luster
   - c. texture
   - d. humus

7. Decomposed animal and plant material is ___.
   - a. soil
   - b. silt
   - c. clay
   - d. humus

Review

Match the definition on the left with the correct term.

8. Soil that is made of tiny grains of rock   **a.** clay

9. Powdery soil that has a slippery texture   **b.** silt

10. Soil that sticks together very well   **c.** sand

Understanding What You Learned

1. What are the five factors that influence soil formation?
2. List the three types of rocks.

Applying What You Learned

1. What are some things people can do to conserve soil?
2. Explain how all animals are dependent on soil.
3. How do the properties of rocks, minerals, and soils help us identify and classify them?

For Your Portfolio

Imagine you are a rock in a garden. Write a book to tell the story of how you were formed, where you have traveled and how, and finally how you ended up in the garden. Draw pictures to illustrate your book. If you wrote another book after this one, what would the title of the next book be?

Unit Review

Concept Review

1. What is gravity?
2. Why might some people listen to the weather report every night?
3. Why does Earth's surface look different than it did 500 years ago?
4. Explain how someone might organize different rocks, minerals, and soils.

Problem Solving

1. Describe what the solar system might be like without gravity.
2. If you were not able to get weather news, how could you predict the weather?
3. Suppose you find a fossil at the edge of a construction site near your home. How could you try to find out the age of the fossil?
4. You are helping your neighbors plant a new garden. They are trying to decide what kinds of plants to put in. Some of the plants they like require a lot of water. Some of the plants they want to put in require very little water. How can you help your neighbors plan their garden?

Something to Do

Take a plastic jar or bottle with a lid (like a peanut butter jar or a soft drink bottle). Make sure the label can be completely taken off so you can clearly see inside. Use different colored sand and make layers in the jar/bottle to make a sand garden. You can be creative and add rocks or leaves to your layers, but make sure that they are placed on the outside wall of the jar/bottle so you can see them when they are covered with layers of sand.

UNIT C

Physical Science

Chapter 1 **Static Electricity and Magnets** **C2**
    Lesson 1: Static Electricity C4
        Activity: Opposites Attract, Likes Repel C10
    Lesson 2: Magnetism C12
        Activity: Magnetic Fields C18
    Chapter Review **C20**

Chapter 2 **Energy Pathways** **C22**
    Lesson 1: Electric Circuits C24
        Activity: Make a Bulb Light Up C32
    Lesson 2: Series and Parallel Circuits C34
        Activity: Making Circuits C40
    Lesson 3: Electromagnetism C42
        Activity: Turning a Magnetic Field
                  On and Off C48
    Chapter Review **C50**

Chapter 3 **Heat** **C52**
    Lesson 1: Heat Production C54
        Activity: Producing Heat C60
    Lesson 2: Heat Transfer C62
        Activity: Observing Heat Transfer
                  by Conduction C68
    Lesson 3: Using Heat C70
        Activity: Using Heat to Do Work C76
    Chapter Review **C78**

Chapter 4 **Matter, Motion, and Machines** **C80**
    Lesson 1: Matter C82
        Activity: Comparing Density C88
    Lesson 2: Motion C90
        Activity: Observing Inertia C96
    Lesson 3: Simple Machines C98
        Activity: Using Simple Machines C106
    Chapter Review **C108**

Unit Review **C110**

CHAPTER 1

Static Electricity and

Ouch! You reach to turn on the faucet and get a shock. It's the first really cold winter day. The heat is on. The air is dry. And you forgot to brush your hand against something before touching the faucet—a simple way to avoid getting a shock. Why was touching the faucet a shocking experience today?

Click. The ends of two magnets snap together. What force is pulling them together? Turn one magnet halfway around, and there is no click. The ends seem to push each other away. The force that pulls two magnets together or seems to push two magnets apart seems amazing. The really amazing thing about that force is that it is related to those little shocks you might get when touching a faucet on a dry day.

The Big IDEA

Electricity and magnets can exert force.

Magnets

CHAPTER SCIENCE INVESTIGATION

Learn about how an electroscope can detect static electricity. Find out how in your *Activity Journal.*

Lesson 1

Static Electricity

Find Out
- How to recognize the effects of static electricity
- How objects become electrically charged
- How electrically charged objects interact

Vocabulary
electrons
protons
electric field
static electricity
grounding

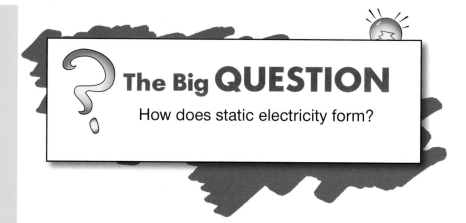

The Big QUESTION

How does static electricity form?

What sometimes makes socks stick together when they come out of a clothes dryer? The invisible force that makes the clothes cling to each other is related to a form of energy you use every day.

Noticing Static Electricity

Electricity is invisible. You can't see it or hear it. But you can see and hear the effects of electricity all around you. Sometimes you can feel the effects of electricity. Do you ever hear clothes crackle and see them cling to each other when you take them out of the clothes dryer? What you see and hear is caused by static electricity.

About 2600 years ago, around 600 B.C., the Greeks noticed the same kind of thing that you have noticed. When a philosopher named Thales (Thā′ leez) rubbed pieces of a yellow stone called amber with sheep's wool, he observed that they did a strange thing—they pulled other small objects to them.

C4

Almost 1000 years later, in A.D. 1570, William Gilbert, an English scientist, also observed that he could make amber beads push away from each other. He could also make a glass bead and an amber bead pull toward each other. From his observations, he inferred that a force was pushing or pulling the beads. He called what he saw *electricity,* after *elektron,* the Greek word meaning "amber."

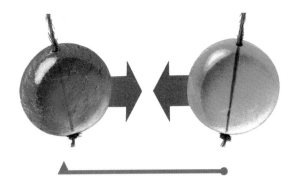

An amber bead and a glass bead pull toward each other.

Atoms and Electrical Charge

To understand why the clothes that come out of a dryer may cling together and why certain beads pull toward or push away from each other, you need to think about tiny pieces of matter that are too small to see. These pieces of matter are called atoms.

All matter is made up of atoms. Within each atom there are even tinier parts, each of which has the smallest possible amount of electricity—one charge. **Electrons** (i lek′ tronz) each have one negative (−) charge. **Protons** each have one positive (+) charge.

Usually, atoms have the same number of protons and electrons. So, the amount of positive (+) and negative (−) electrical charge is the same. When the atoms that make up two objects have the same amount of positive and negative electrical charge, we say they are electrically neutral.

Electrically neutral objects don't push away from or pull toward each other. It's only when objects become electrically charged—when they have extra positive or negative charges—that they act in interesting ways.

An atom's electrons each have one negative (−) charge. Its protons each have one positive (+) charge.

Electron

Proton

Illustration not to scale

Electrons Move

How do items like socks and beads become electrically charged? If you think about the socks that you take out of a dryer and Thales's amber, you'll realize that they have one important thing in common. In both cases, the objects were rubbed by objects made of a different substance. As the socks tumbled around in the dryer, they bumped and rubbed against sweatshirts, pants, and towels. The amber was also rubbed with a piece of wool.

In some materials, electrons are easily pulled off atoms. This means that the electrons, which have negative (–) charges, can move to another material if they rub against it.

The electrons in wool can fairly easily move to atoms in some other materials. If a balloon is rubbed with a piece of wool, some electrons will move from the wool to the balloon. The balloon will gain electrons and become negatively charged. Only electrons can move. Protons cannot move. Because it will lose electrons, the piece of wool will have more protons than electrons. It will be positively charged.

Rubbing a plastic comb through hair can leave both substances electrically charged.

Here is another example of how objects become electrically charged. Look at the girl in the picture. Her hair is standing out from her head because it has become electrically charged. When the girl combs her hair, the electrons move from her hair to the comb. The balance between protons and electrons in her hair and in the comb is upset. Both objects become charged.

If both items have the same electrical charge, they will push away from each other. The like charges repel each other. But if the two items have opposite electrical charges, they will pull toward each other. Opposite charges attract. Look at the girl again. Do the comb and the hair have like charges or opposite charges?

Charges can attract or repel without the two items even touching. This happens because of electric fields. An **electric field** is the space around a charged object that pushes or pulls on another charged object. Electric fields are the reason why your socks sometimes get "static cling" or why your hair may stand up when it rubs against your coat hood.

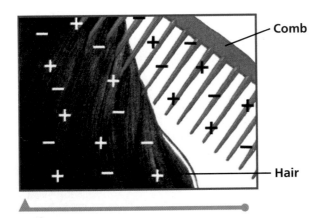

Before combing, the hair and the comb are neutral.

After combing, electrons have moved from the hair to the comb.

Effects of Static Electricity

The kind of electricity that makes clothes cling is called **static electricity.** The word *static* means "not moving," so the electrical charge is on something and not moving (even though it moved to get there). This makes it different from the electricity that runs lights, computers, and many other electrical devices. The electricity that we use every day flows continuously, but static electricity does not.

Whenever you shuffle across a carpet and then touch a metal doorknob and get a slight shock, you are feeling the effects of static electricity. Electrons move from the carpet to your shoes when the materials rub together. The electrons travel through your body and are then discharged into the doorknob as you touch it.

The small amounts of static electricity you've experienced in the form of shocks like these are annoying but harmless to you. Sometimes, though, huge amounts of charge build up on objects. Static electricity builds up in clouds when atoms bump against each other. When that electricity travels to the ground, it makes a giant spark we call lightning. Big discharges of static electricity—lightning—can give a very dangerous shock that can kill people or start fires.

A lightning rod grounds the enormous discharge of static electricity of a lightning bolt.

Static Electricity Around Us

A lightning rod is a device that is put on the roof of a building to carry the electricity safely to the ground. Part of the reason this works is that the rod is connected to the ground with a wire. Instead of traveling through the building, the lightning travels down the path created by the rod and the wire, so it discharges harmlessly into the ground. Getting rid of extra electrical charge by transferring it to the ground in this way is called **grounding.** Grounding is the easiest way to get rid of static charge.

A photocopier uses static electricity.

A device that uses static electricity is a photocopier. Many photocopiers work by putting an image on a roller inside the machine. The image is recorded on the roller as invisible positive charges. These charges attract a black powder so that the image can be seen. The image is then moved to electrically charged paper. Heat melts the black powder onto the paper so the photocopy can be used.

CHECKPOINT
1. What are the effects of static electricity?
2. How do objects become electrically charged?
3. How do electrically charged objects interact?

 How does static electricity form?

ACTIVITY

Opposites Attract, Likes Repel

Find Out
Do this activity to see how objects can become electrically charged and how charged objects interact with other objects.

Process Skills
Observing
Communicating
Inferring
Experimenting
Predicting

What You Need

small bits of paper

two balloons

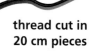
thread cut in 20 cm pieces

Activity Journal

piece of wool cloth

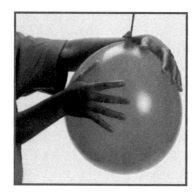

What to Do

1. Working in groups of two, blow up two balloons (balloon 1 and balloon 2) and tie them to separate pieces of thread.

2. Bring balloon 1 near balloon 2. **Observe** how the balloons interact with each other. Write your observations.

3. Now, electrically charge both balloons by rubbing them several times with the wool cloth.

4. Bring balloon 1 toward balloon 2. Repeat this several times. Write your observations. Then write why you think the balloons behaved as they did.

5. Experiment further with the two balloons. Predict what will happen when you bring the wool cloth near the balloons. Observe what happens when you move each balloon near the bits of paper. What happens every time you move the balloons near your hair or near the wall? Write your observations. Try to explain why the objects acted the way they did.

CONCLUSIONS

1. Did the balloons act the same every time? If not, why do you think they sometimes acted differently?
2. Think about the objects that moved. Did you have to touch the objects together to make them move? Why?

ASKING NEW QUESTIONS

1. How did you make electrons move from one type of matter to another in this activity?
2. Thales and William Gilbert inferred that static electricity existed, because of their observations. List one observation and one inference you did in this activity.

SCIENTIFIC METHODS SELF CHECK

✔ Did I **observe** and **write** what happened to the balloons and the other objects?

✔ Did I **infer** the causes of the interactions I observed?

✔ Did I **experiment** with the balloons?

Lesson 2

Magnetism

Find Out
- What a compass can do and why it is useful
- How magnets interact
- What a magnetic field can do

Vocabulary
magnetic poles
magnetic field
magnetized

The Big QUESTION
How can you magnetize an object for a short time?

You can't see exactly what happens when one magnet makes another move without touching it or when two magnets stick together. In this way, magnetism is a lot like static electricity. The force that is at work is invisible, but its effects definitely are not.

Discovering Compasses

Around 300 B.C. a Chinese scholar noticed that pieces of a certain type of rock would always move into a certain position if they were allowed to swing freely. After investigating further, others in China found that this rock could be used to find direction. The rock always moved in a line from north to south.

By A.D. 1000 this discovery had spread, and Chinese sailors were using pieces of the rock to help them find their way at sea. They either hung pieces of the rock from string or put them on wooden floats in pans of water.

The idea of using these rocks to find direction eventually reached the Arab world and then Europe. In Europe, pieces of the rock came to be called *lodestones* (lōd′ stōnz), which means "leading stones." Because sailors could then tell direction without looking at the sun and stars, they could find their way on a completely cloudy day or night. Lodestones led them when they needed to know direction.

Magnetism is all around us. It is helping these hikers find their way back to their campsite.

Compasses and Magnets

Magnetite is the name by which we know the mineral today. As you may have guessed, it is a natural magnet. A magnet is a material or device that attracts iron and some other metals. Magnets also have the property of lining up in a north-south direction when allowed to swing freely.

It is easy to make a simple compass and re-create what the Chinese discovered. Suppose you made a sling for a bar magnet and waited for it to stop moving. In what directions would the ends point? You could use a small compass to check if the ends are lining up in a north-south direction.

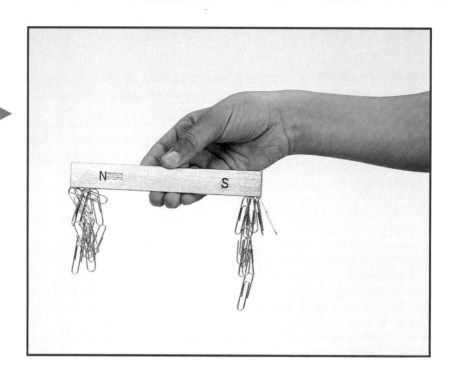

More paper clips are clumped at the poles, where the pull of the magnet is strongest.

The Push and Pull of Magnets

Magnetic Poles

A bar magnet has two poles. They are marked N, which means "north-seeking magnetic pole," and S, which means "south-seeking magnetic pole." **Magnetic poles** (mag net′ ik pōlz) are the places on the magnet where the attraction is the strongest. When a bar magnet is suspended to act as a compass, its poles pull toward the magnetic poles of Earth, which itself acts as a giant magnet. The poles of a bar magnet can attract paper clips, iron filings, steel balls, and other items that contain iron.

You can feel the attracting and repelling forces of a magnet when you move the poles of two different magnets toward each other. The like poles (N-N or S-S) push away from, or repel, each other. The opposite poles (N-S) attract each other. You can also see how magnetic poles act when iron filings are placed around them.

Look at the first pair of magnets. Opposite poles are near each other (N-S), and there are lots of iron filings being pulled into the space between them. The filings are being pulled by the forces of attraction that exist between opposite poles.

Now, look at the second pair of magnets. Notice that when the like poles are near each other, the filings are pushed out to the sides. You can see them being pushed by the repelling forces that exist between like poles.

Magnetic Fields

A **magnetic field** is the space around a magnet that pushes or pulls on iron-containing objects, other magnets, and some other materials. The lines made by iron filings can show you where the magnetic field is. Although the patterns of lines around differently shaped magnets look different, they always have something in common. The lines always curve out from one pole, around, and back to the other pole.

Magnets don't have to touch iron-containing objects or other magnets to make them move. You can see this from looking at the magnetic fields on this page, and you can experience it when you handle magnets.

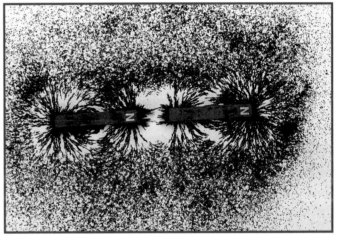

Opposite poles are pulling the iron filings between them.

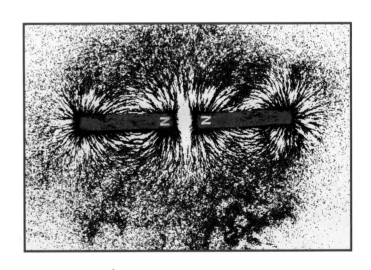

Like poles are pushing the iron filings out.

Making a Magnet

Iron filings are attracted to magnets because each filing becomes a small magnet when it is placed in a magnetic field. This is why magnets attract iron and things that contain iron, such as steel. Each little piece of iron becomes **magnetized,** or made into a magnet.

A nail or other iron-containing object can also be magnetized so that it acts as a temporary magnet. The picture shows how.

To magnetize the nail, you pull it over one pole of a bar magnet ten times or more in the same direction. The magnetic properties of the nail will become a little stronger with each pull. With enough pulls, the magnetized nail can be made strong enough to pick up several paper clips, although it will not be as strong as a bar magnet.

The magnetized nail is different from a bar magnet in another way. Its magnetic power will become weak after a few hours. Slowly, it will lose its ability to attract iron-containing materials.

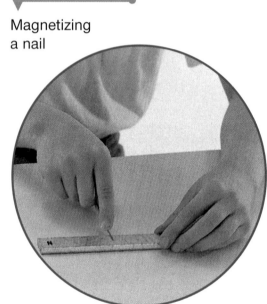

Magnetizing a nail

Ceramic magnets are used in stereo speakers.

Other Magnets

Besides iron and steel, there are other materials that can be magnetized. Ceramic magnets are made from metals that have been combined with oxygen. They are brittle and can break if you drop them. They weigh less than iron magnets. They can be found in the speakers of your stereo or CD player. There are even magnets made of many very small magnet pieces stuck in flexible plastic. The plastic makes these magnets very easy to bend and cut. You may have some of these magnets on your refrigerator at home.

Magnets can be found in many places in businesses, homes, and schools. Because magnetic fields can pull or push objects without touching them, magnets can be used to move objects like scrap metal. Magnetic forces can even work through other materials, such as glass and paper. That is why paper notes can be held on a message board with a magnet. Can you find other examples of magnets in your classroom?

Magnetic forces can even work through paper.

CHECKPOINT

1. Describe what a compass can do and why it is useful.
2. Explain the two main ways that magnets can interact.
3. Explain what a magnetic field can do.

 How can you magnetize an object for a short time?

ACTIVITY

Magnetic Fields

Find Out
Do this activity to see how a magnetic field can attract and repel.

Process Skills
Experimenting
Communicating
Observing
Inferring

What You Need

two bar magnets

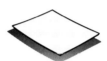
piece of white paper

safety goggles

one resealable bag with two spoonfuls of iron filings in it

Activity Journal

What to Do

1. Put a magnet on a desk. **Try** to move the magnet across your desk without touching it. **Record** how you can make the magnet move. **Draw** the combinations.

2. Put a magnet flat on your desk with a piece of paper on top of it.

   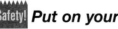 **Put on your safety goggles.**

3. Place some of the iron filings on the paper above the magnet. **Sketch** what you see.

4. Now, put the two magnets on your desk and place the ends about 2 cm apart. **Write down** which two ends are facing each other. Put the paper on top of the two magnets and place some iron filings on the paper around the two magnets. **Sketch** what you see.

5. Repeat step 4 with different ends of the magnets together. **Label** which ends are together in your **sketches**.

CONCLUSIONS

1. Which ends of the magnets pulled toward each other? Which ends pushed away?
2. Where did most of the filings clump together? What does this show about the magnet?
3. Which parts of the magnet would be best for picking up paper clips?

ASKING NEW QUESTIONS

1. Look at your sketches of the filings. How do you think the magnets pushed or pulled each other without touching?
2. Suppose you have one bar magnet with marked poles and one with unmarked poles. How could you find out which pole on the unmarked magnet is *N*?

SCIENTIFIC METHODS SELF CHECK

✔ Did I **experiment** with different ways to move the magnet without touching it?

✔ Did I **observe** and **sketch** patterns that the iron filings formed?

✔ Did I **infer** how these forces and patterns are related to magnetic poles and magnetic fields?

Review

Reviewing Vocabulary and Concepts

Write the letter of the answer that completes each sentence.

1. ___ each have one negative (−) charge.
   a. Magnetic charges   b. Protons
   c. Atoms   d. Electrons

2. ___ each have one positive (+) charge.
   a. Atoms   b. Magnetic charges
   c. Protons   d. Electrons

3. The shock you feel when you touch a doorknob in winter is due to a build up of ___.
   a. ground   b. static electricity
   c. magnetism   d. neutral atoms

4. The easiest way to get rid of static electricity is ___.
   a. grounding   b. lightning
   c. negative charge   d. positive charge

5. The places on a magnet where the attraction is strongest are the ___.
   a. electrons   b. magnetic poles
   c. electrical fields d. protons

Match each definition on the left with the correct term.

6. the area around a charged object that pushes and pulls objects
7. getting rid of excess electrical charge
8. what makes socks sometimes stick together
9. the area around a magnet that pushes and pulls objects
10. what an object is when it has been turned into a magnet

a. static electricity
b. magnetized
c. magnetic field
d. electric field
e. grounding

Understanding What You Learned

1. When socks rub against other clothes in a dryer, they can get static cling. What moves from atoms of a sock to atoms of other clothes to cause this?

2. When a balloon is rubbed with wool, the wool sticks to it. What causes the wool to stick?

3. If someone puts a rock in your hand and tells you that it has a lot of magnetite in it, what do you think the rock will be able to do?

4. What is within a metal object that is attracted by a magnet?

5. List two things that have magnetic or electrically charged parts that are often found in homes and schools.

Applying What You Learned

1. Suppose you have a nail and a magnet. Explain how you can turn the nail into a magnet.

2. What might a hiker use to keep track of his or her way? How would this help?

3. If you place the ends of two bar magnets together and they push against each other, what do you know? What if the ends pull together?

4. Explain one method used to stop dangerous electrical discharges from damaging buildings.

 5. Explain how static electricity and magnets are similar.

For Your Portfolio

Think about what you have learned about magnets and static electricity. Computers are very sensitive to electric and magnetic fields. Make a plan for testing items that you want to place near a computer to make sure they will not damage the computer. Write directions for your test so that others will know what to do.

CHAPTER 2
Energy PATHWAYS

What do a video game, a lightning storm, and an elevator have in common? Electrical energy. You know that lightning is the discharge of static electricity, but how does electrical energy light things up or make things move?

Power lines carry electric current from power plants to your home. When you turn on a television, you use electrical energy from power plants. These power plants provide electrical energy that can be converted to heat, light, or motion.

In the past, people used their own energy to climb stairs, and they used horses to get around. Today, we rely on electric current to run elevators, some trains, and even some cars. Electrical energy has made our lives very different.

The Big IDEA

Electric current moves through different circuits.

CHAPTER SCIENCE INVESTIGATION

Learn how electric current moves through different materials. Find out how in your *Activity Journal.*

Lesson 1

Electric Circuits

Find Out
- What an electric circuit is
- How switches control electrical energy
- Which materials conduct and insulate electric current

Vocabulary
electric current
circuit
switch
electric conductors
insulators
resistors

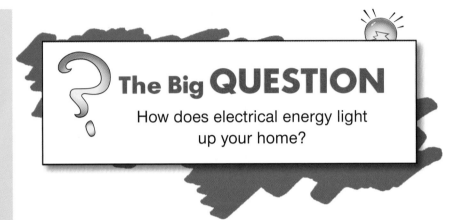

The Big QUESTION

How does electrical energy light up your home?

Imagine what the world was like before the discovery of electricity. There were no TVs, radios, or CD players. There were no electric lights to help people see in the evening. There were no electric refrigerators to keep food cold or electric furnaces to keep people warm in the winter.

The Flow of Electricity

Each day you turn lights on and off in your home. How does electrical energy get to the lightbulbs? As you recall, static electricity involves the movement of electrons. When the electrons move, they produce positively or negatively charged atoms that push or pull against other charged atoms.

Electrons can also move along a path. The flow of electrons along a path is called an **electric current.** To see how electrons flow, look at the marbles in the photo. If you push one marble, the others will also move. If you continue pushing one of the marbles, all of the marbles will flow around the glass. The path of the marbles is like an electric current.

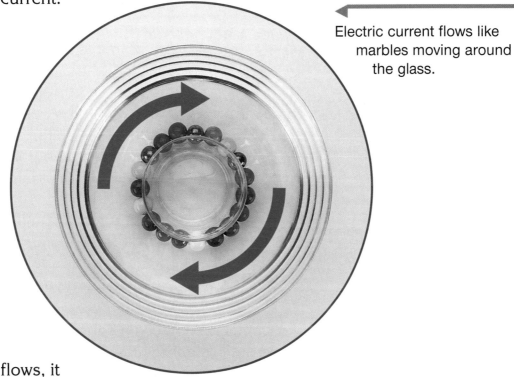

Electric current flows like marbles moving around the glass.

When electric current flows, it needs something to move through. It usually moves through wires. These wires form a **circuit**, which is a path for electrical energy to move through. When a circuit is complete, it allows the electric current to flow.

The train track is a complete pathway for a train to follow. A circuit is a complete pathway for electric current to follow.

Open and Closed Circuits

The girl has opened the train circuit by "cutting" the track. Electric current can't flow when a circuit is open.

If the train track is broken in one section, the train cannot run around it. The same is true with electric current. A pathway with a break in it is called an open circuit. If the circuit is open, the electric current cannot move along it. A pathway with no breaks is called a closed circuit. Electric current can flow only if the circuit is closed. How can you make the train run again if the track is broken? You can put the track together again, making a closed circuit so the train can run.

A circuit also needs a source of energy. One source of energy is a battery. The electric current moves from the source, around the circuit, and then back to the source. The picture shows a complete circuit. The battery is the source of electrical energy. Electric current flows from the battery through the circuit. The wires connect the battery to the lightbulb. When the circuit is closed, the lightbulb turns on. When it is open, the lightbulb turns off.

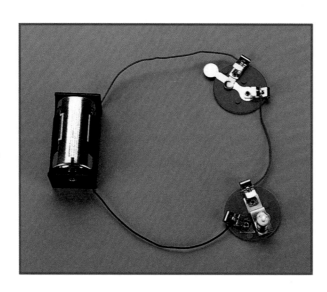

The source of electrical energy in this circuit is the battery.

C26

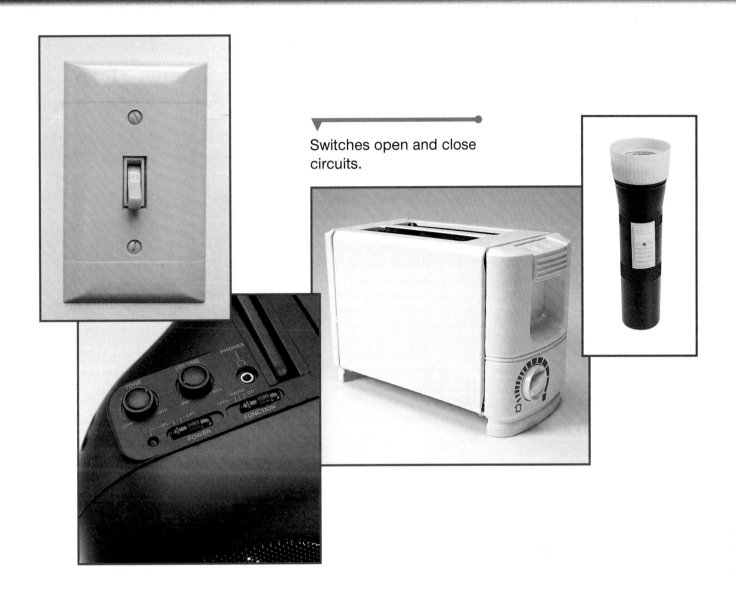

Switches open and close circuits.

The source of energy for the lights in your home is a power plant. The electric current moves from the power plant along the power lines to your home. When the circuits in your home are closed, the lights turn on.

Sometimes, you need to open circuits, as you do when you want to turn off the lights or to stop an electric train. The easiest way to do this is to use a switch. A **switch** is a device that is used to open and close circuits without disconnecting wires or unscrewing bulbs. Switches also help save energy because you can open the circuit and stop the current with the switch. Switches can be found on most electrical devices.

Conductors and Insulators

Electric current moves through some materials more easily than it does through others. Materials that electric current flows through easily are called **electric conductors** (kon duk′ tərs). Most metals such as gold, silver, copper, and aluminum are good conductors. Other common things, such as pencil lead, also conduct electric current.

Materials that electric current does not flow through easily are called **insulators** (in′ sə lā tərs). Air, rubber, and plastic materials are insulators. In an electrical wire, the metal wire is inside of a plastic covering. The metal wire carries the electric current into an appliance. The plastic covering acts as an insulator by keeping the electric current from flowing where it should not go. The plastic covering also makes the wire safe to touch. If the plastic is cracked or damaged, the wire is not safe to touch. This is because your body can also conduct electricity. Electric current can flow through your body and burn you.

Glass, ceramics, and fiberglass are insulators. They can be used to make safety materials like this ladder to stop people from coming in contact with electric current.

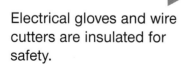

Electrical gloves and wire cutters are insulated for safety.

Resistors

Some materials resist, but don't stop, electric current. They are called **resistors** (rē sis′ tərz). When electric current flows through these materials, they get hot because resistors change some electrical energy to heat. The wire inside a lightbulb is a resistor. When the electric current flows, the wire inside the bulb gets hot enough to produce a bright glow. In the summer, lights that are turned on can make a room feel too warm. You need to be careful not to touch a lightbulb that is on, because you could burn yourself.

The burner on an electric stove is also a resistor. When electric current flows through the burner, the burner begins to heat up and glow. This heat can be used for cooking.

The burner on an electric stove is a resistor.

Circuit Diagrams

People who build and design electric circuits draw pictures of the wires, sources of energy, switches, and the electrical devices. They use diagrams with symbols that show the parts of the circuit. Diagrams can show what is in a circuit and how it works. Diagrams can help technicians follow the path of electrical energy in households, businesses, or machines.

The symbols in an electrical diagram show parts of a circuit. Look at the circuit diagram and see what symbols match up with the chart. Use your finger to trace the flow of electric current in the diagram.

Symbols for Electric Circuits		
Item	Purpose	Symbol
Wire	Conducts electric current	——
Battery	Stores and supplies electricity	+⊣⊢
Lightbulb	Provides electric light	-⟋⟍-
Switch	Completes or breaks a circuit	-o ⟋ o-

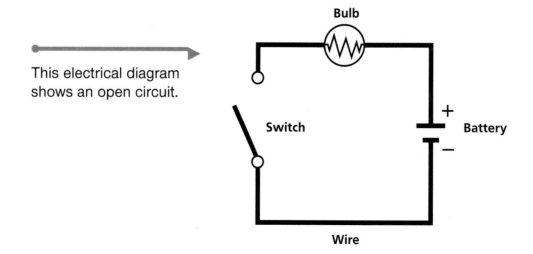

This electrical diagram shows an open circuit.

C30

Can you draw an electrical diagram of this photo?

You use electrical energy to make things happen every day. In a lightbulb, electrical energy is changed into light. When you make toast, electrical energy in the toaster becomes heat. When you turn on a fan, electrical energy drives the motor that moves the fan blades. Think about how many different ways you use electrical energy and electric circuits each day.

CHECKPOINT

1. What is an electric circuit?
2. How do switches control electrical energy?
3. What materials conduct electric current?
 How does electrical energy light up your home?

ACTIVITY
Make a Bulb Light Up

Find Out
Do this activity to learn how electrical energy can operate lights.

Process Skills
Predicting
Experimenting
Communicating
Defining Operationally

What You Need

D-cell battery

aluminum foil

flashlight bulb

scissors

two 20 cm pieces of masking tape

Activity Journal

What to Do

1. First, make a wire by sticking the tape onto the aluminum foil.
2. Cut around the tape. Make sure to leave the foil on each end a little longer than the tape.
3. Fold the tape in half lengthwise with the foil on the inside. You now have an aluminum wire with an insulator.
4. Make another wire with the other piece of tape and some more aluminum foil.

5. **Predict** what will happen when you connect the wires to the bulb and the battery. **Write** your prediction.

6. Now, **find** as many ways as you can to light up the flashlight bulb by using your wires and the battery. Repeat each plan several times to test it.

   Safety! Do not connect both ends of the battery with one wire.

7. **Draw** each arrangement you try—even if the arrangement doesn't work. Be sure to **mark** the ones that do work. Add arrows in your drawings to show the direction of the electric current.

Conclusions

1. Compare your predictions with your observations.
2. What does the battery do in your investigation?
3. What did you do each time you made the bulb light up?

Asking New Questions

1. Why did you fold the aluminum foil on the inside of the tape?
2. Do you need two wires to make the bulb light up? Explain.

SCIENTIFIC METHODS SELF CHECK

✔ Did I **predict** what would happen when I connected the bulb to the battery with the wires?

✔ Did I **experiment** by touching the wires to different parts of the bulb and the battery?

✔ Did I **record** my observations?

✔ Did I **define** when I made a closed circuit?

Series and Parallel Circuits

Find Out
- What a series circuit is
- What a parallel circuit is
- How electric current can be controlled

Vocabulary
series circuit
parallel circuit
fuses
circuit breakers
short circuit

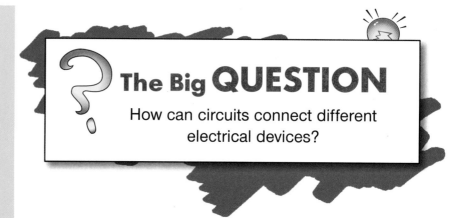

The Big QUESTION

How can circuits connect different electrical devices?

*T*here is usually more than one way to get something done. That is also true of electric circuits. There are many ways to connect appliances and turn them on and off. Think about the lights along a highway or on your street. How are they all connected?

Series Circuits

Every time you switch on a flashlight, you have completed a simple circuit. But circuits can also be more complex. Lights and appliances can be connected by two kinds of circuits: series circuits and parallel circuits.

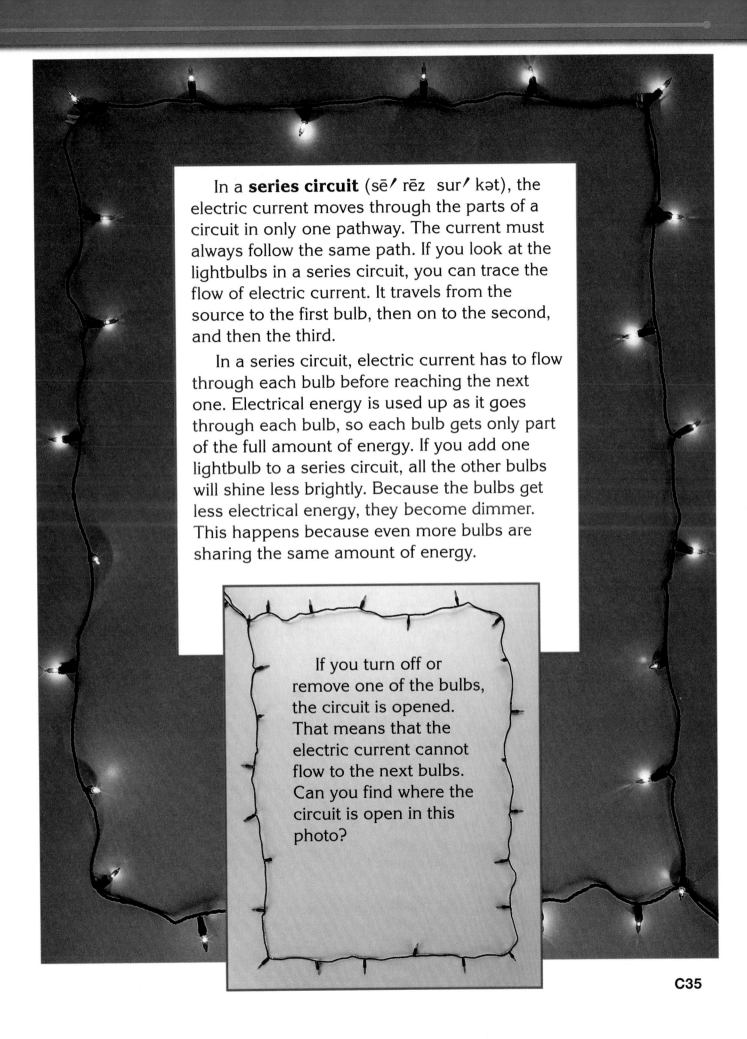

In a **series circuit** (sē/ rēz sur/ kət), the electric current moves through the parts of a circuit in only one pathway. The current must always follow the same path. If you look at the lightbulbs in a series circuit, you can trace the flow of electric current. It travels from the source to the first bulb, then on to the second, and then the third.

In a series circuit, electric current has to flow through each bulb before reaching the next one. Electrical energy is used up as it goes through each bulb, so each bulb gets only part of the full amount of energy. If you add one lightbulb to a series circuit, all the other bulbs will shine less brightly. Because the bulbs get less electrical energy, they become dimmer. This happens because even more bulbs are sharing the same amount of energy.

If you turn off or remove one of the bulbs, the circuit is opened. That means that the electric current cannot flow to the next bulbs. Can you find where the circuit is open in this photo?

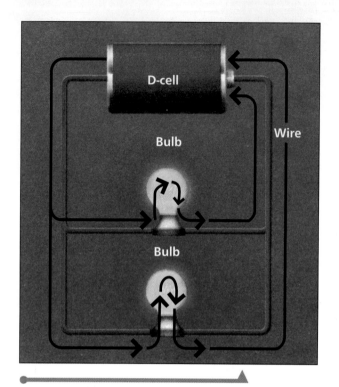

This circuit shows two lightbulbs connected in parallel.

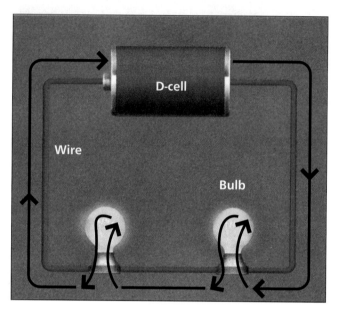

This circuit shows two lightbulbs connected in series.

Parallel Circuits

In a **parallel circuit** (par′ ə lel′ sur′ kət), the electric current flows through different parts of the circuit at the same time. It's like having a detour you can take when the road is blocked. In a parallel circuit, if one lightbulb burns out, all the other bulbs and appliances do not go out. Each bulb has its own separate pathway for the current to flow through. So, removing one bulb does not open the circuit.

In a parallel circuit, the electrical energy only flows where the switches are closed. Different switches can turn different appliances on and off in a parallel circuit.

With parallel circuits, the motor in the refrigerator in this restaurant can continue to run even though the lights are turned off.

Series or Parallel Circuits?

Why are homes and buildings wired with parallel circuits instead of series circuits? Think about restaurants or other businesses that rely on refrigerators or freezers. What would happen to these appliances if they were connected in a series with the switch that turns the lights off? Wouldn't they go off too? They would. Putting these appliances on a parallel circuit makes it possible for them to continue to run when the lights are turned off for the day.

Look at the photo of the streetlights. These lights must be connected in a parallel circuit. If one of the lights burns out, the rest of the lights will still light up the road for passersby.

Streetlights are connected in parallel circuits.

Controlling Electric Current

Fuses and Circuit Breakers

It is often necessary to limit the amount of electric current that flows through wires. Too much current is dangerous because it heats the wires. The hot wires can start fires. A switch only controls whether or not the current can flow by opening and closing a circuit. There are other devices that limit how much electric current moves along the wires. **Fuses** and **circuit breakers** act as safety switches. They open the circuit to prevent too much electrical current from flowing through wires.

A fuse has a very thin strip of metal in it. The strip melts when too much electricity flows through it. This opens the circuit. Fuses can be used only once.

When too much electricity flows through a circuit breaker, a piece of metal gets hot and expands to push the switch open. Circuit breakers can be reset and used again.

Fuses and circuit breakers automatically shut off the flow of electric current when too much current begins to flow through a circuit. This usually happens when there are too many appliances and lights on one parallel circuit. If some lights or appliances are shut off and the fuse is replaced, the circuit is closed again. Then, the electric current will flow again.

Short Circuits

Sometimes, wires get broken or get moved so that they touch each other. When this happens, the electric current travels through the circuit without passing through any device. This is called a **short circuit.** Because no device is in the circuit to use up the electrical energy, too much current flows. As a safety measure, circuit breakers open the circuit to prevent a fire. To fix the circuit, the broken wires must be replaced.

Imagine a world without electric current flowing through circuits. You wouldn't be able to run the computer or the TV in your home. Understanding how electrical energy flows through different kinds of circuits gives you a good idea of how the lights and electrical appliances in your home work.

CHECKPOINT

1. What is a series circuit?
2. What is a parallel circuit?
3. How can you control the flow of electrical energy?

 How can circuits connect different electrical devices?

ACTIVITY

Making Circuits

Find Out
Do this activity to learn how to make different electric circuits.

Process Skills
Experimenting
Communicating
Predicting
Observing
Defining Operationally

What You Need

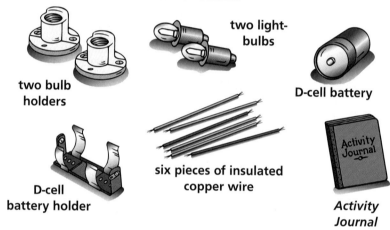

two bulb holders

two lightbulbs

D-cell battery

D-cell battery holder

six pieces of insulated copper wire

Activity Journal

What to Do

Safety! *Never experiment with the electric circuits in your home. The source of energy is very large and the electric current is very dangerous. The D-cell battery you will use here is a much smaller source of electrical energy, so it is safe.*

1. Make a circuit using the wires to connect the D-cell battery to two bulbs. **Experiment**, using the fewest wires you can. **Test** your circuit several times to see if both bulbs light up.

2. Use electrical symbols to **draw** your circuit diagram. Label it "Circuit 1."

3. **Predict** what will happen when you remove one of the bulbs.

4. **Test** your prediction. Remove one bulb. **Observe** what happens. Have you constructed a series or a parallel circuit? How do you know?

5. **Record** your observations.
6. Now make the other type of circuit. Use the fewest number of wires you can.
7. **Draw** your circuit diagram. Label it "Circuit 2."
8. **Predict** what will happen when you remove one of the bulbs. **Test** your prediction several times. **Observe** and **record** the results for each test.

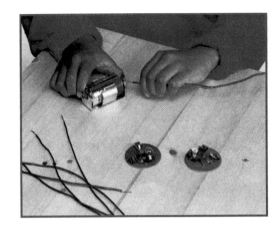

Conclusions

1. Compare your predictions with your observations.
2. What happened in each circuit when you removed a lightbulb?
3. Which circuit had the brighter bulbs? Why?

Asking New Questions

1. How is removing and replacing a bulb like opening and closing a switch?
2. Which kind of circuit do you think works better to light your home? Why?

SCIENTIFIC METHODS SELF CHECK
- ✔ Did I **experiment** with many different ways to complete the circuit?
- ✔ Did I **record** my observations?
- ✔ Did I **predict** what would happen?
- ✔ Did I **distinguish** between a series and a parallel circuit?

LESSON 3

Electromagnetism

Find Out
- How you can magnetize a nail
- How to turn an electromagnet off
- How we use electromagnets

Vocabulary
electromagnet

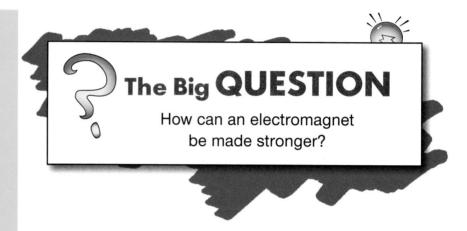

The Big QUESTION

How can an electromagnet be made stronger?

Electric current can be used to make a magnet. Many electrical machines and tools around you use this kind of magnet because it can be turned on and off. Most of these machines—such as VCRs, TVs, and even telephones—run because of magnets and electricity.

Magnetizing Objects

Most things are not magnets. A nail alone isn't much of a magnet. But if you pull the nail over a magnet, the nail becomes a temporary magnet. A straight wire with current flowing through it is a weak magnet. Whenever electric current passes through a wire, a magnetic field is created. But if you wrap a wire around and around into a coil and pass a current through it, the wire becomes a stronger magnet. The coil shape concentrates the magnetic field into a small space.

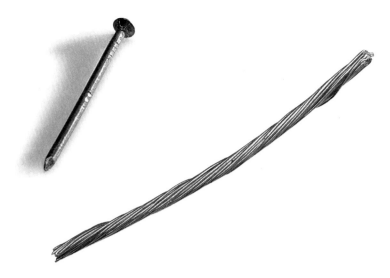

A nail or a piece of wire is not a magnet.

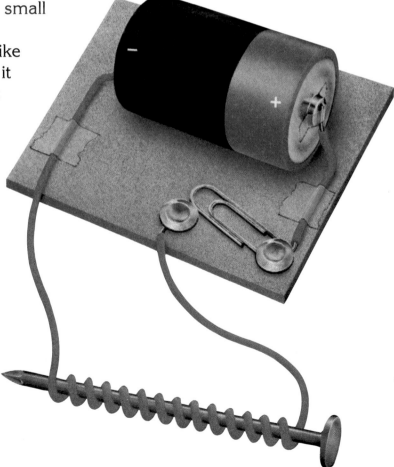

The electric current from the battery flows through the wire, making the nail a strong magnet. The more turns of the wire, the stronger the magnet.

Putting a piece of iron inside the coil of wire makes the magnet even stronger. The iron concentrates the magnetic field produced by the current even more. While the electric current is flowing, the magnet can pick up small iron objects. When the current is turned off, the iron stops acting like a magnet and drops any objects it was holding. An **electromagnet** (ə lek′ trō mag′ nit) is an arrangement of iron wrapped in a coil of wire carrying an electric current. An electromagnet has north and south poles, like a permanent magnet.

An electromagnet can be made even stronger by increasing the number of coils. A nail with wire wrapped around it ten times will be a stronger electromagnet than a nail with only five wraps.

Turning Electromagnets On and Off

"Seeing" a Magnetic Field

Magnetic fields are invisible, but you can notice the effects of those fields. You can "see" the magnetic field of an electromagnet by using a compass. Look at the nail wrapped in the coils of wire. When the electric current flows through the wire, the compasses all point toward the electromagnet. The current flowing in the wire is producing a magnetic field.

If you move the electromagnet, the compass needles will turn in the direction of the electromagnet. The magnetic field is moving, and the needles show the moving field.

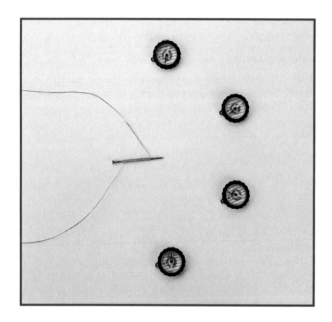

All the compass needles are pointing toward the electromagnet because current is flowing in the wire.

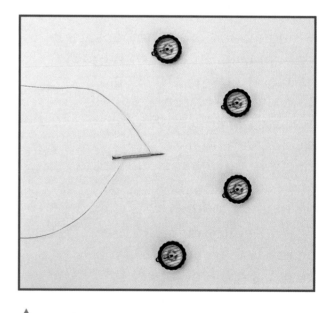

When the electromagnet is turned off, the compass needles no longer point to it.

Electromagnets can move hundreds of kilograms of scrap iron.

Magnets That Use Electric Current

Electromagnets come in all sizes, from little ones in some doorbells to big ones that can lift scrap steel from cars and trucks. Some scrap yards use giant cranes with huge electromagnets to lift and separate waste iron from other materials. When the switch is turned on, electric current flows through the coils. The electromagnets can then pick up steel or iron—even a car. When the switch is turned off, the circuit is open and the electric current stops flowing. The electromagnets inside lose their magnetism, and they no longer attract iron.

Other Uses of Electromagnets

Electromagnets can help produce pictures and sound. The surface of videotapes and audiotapes is magnetic. Sounds and pictures are recorded on the tape as tiny areas of different magnetism. When the tape is run through a player, the areas pass near another weak electromagnet called a head. As they pass, the magnetic areas cause the magnetic field of the head to change. These changes are seen as pictures on a TV screen or heard as sounds from a speaker.

Many machines around you run because of electromagnets. They can ring doorbells and open door latches. When you push the button on an electric doorbell, a switch inside of the house makes a current flow and turns on the electromagnet. An iron bar that is attached to a hammer is attracted to the electromagnet. When the bar moves, the hammer strikes a bell. The circuit is then broken. The electromagnet releases the bar and hammer. This is repeated many times very quickly so you can hear a ringing sound.

An electromagnet causes the hammer to hit the bell, making the doorbell ring.

An electric doorbell rings when current flows through the electromagnet.

The archway of a metal detector contains wire coils with electric current flowing through.

Electromagnets have many useful roles in our everyday lives. Metal detectors, such as those seen in airports, use electromagnets. Electric motors and electric generators depend on the flow of electric current. These motors use electromagnets to turn electricity into movement. Motors run toy trains, washing machines, fans, elevators, and the windshield wipers on cars. Can you imagine what your daily life would be like without electromagnets?

CHECKPOINT

1. How can you magnetize a nail?
2. How can you turn an electromagnet off?
3. Name two ways we use electromagnets.
 How can an electromagnet be made stronger?

ACTIVITY

Turning a Magnetic Field On and Off

Find Out
Do this activity to learn how a magnet can be turned on and off and how a magnet can be made stronger.

Process Skills
Observing
Defining Operationally
Communicating
Predicting
Experimenting

What You Need

two steel nails

1 m of insulated copper wire

small metal paper clips

D-cell battery holder

D-cell battery

Activity Journal

What to Do

1. Touch each nail to the paper clips to **see** if they will stick. **Record** your observations.

2. Leaving about 20 cm of extra wire at the beginning, wrap 15 loops of wire tightly around one nail. Be careful not to overlap any loops. The coils in the wire must be very close together.

3. Connect one end of the wire to the battery holder. Have a partner hold the other end of the wire to the other end of the battery holder while you carefully move the nail into the pile of paper clips.

   Safety! *Do not leave the battery connected for more than a few seconds. You will need to work quickly.*

4. Pick up as many paper clips as you can. Move the nail away from the pile, and then take the end of the wire away from the battery holder.
5. **Record** the number of paper clips you picked up.
6. **Predict** what you have to do to pick up more paper clips. **Test** your prediction.
7. **Graph** the results of your different tests.

Conclusions

1. What happened when you placed the nail in the paper clips when the circuit was closed?
2. What happened when the circuit was opened?
3. What did you do to pick up more paper clips?

Asking New Questions

1. What other items could the nail pick up?
2. What kind of machine could you make with your electromagnet?

> **SCIENTIFIC METHODS SELF CHECK**
> ✔ Did I **observe** how paper clips stick to either nail?
> ✔ Did I **predict** and **experiment** with ways to pick up more paper clips?
> ✔ Did I **define** which nail was magnetized?
> ✔ Did I **record** my observations?

Review

Reviewing Vocabulary and Concepts

Write the letter of the answer that completes each sentence.

1. To open a circuit, you can use a ___.
   - **a.** resistor
   - **b.** switch
   - **c.** conductor
   - **d.** battery

2. Gold, silver, copper, and aluminum are all examples of ___.
   - **a.** insulators
   - **b.** electric circuits
   - **c.** resistors
   - **d.** electric conductors

3. If a string of holiday lights does not come on when one bulb is burned out, it is an example of a ___.
   - **a.** circuit breaker
   - **b.** series circuit
   - **c.** insulator
   - **d.** parallel circuit

4. The lights in your home are connected by ___.
   - **a.** parallel circuits
   - **b.** a circuit breaker panel
   - **c.** electrical energy
   - **d.** electric currents

5. ___ can lift tons of scrap steel.
   - **a.** Resistors
   - **b.** Series circuits
   - **c.** Electromagnets
   - **d.** Switches

Match each definition on the left with the correct term.

6. wires that form a path for electrical energy to flow through
7. a circuit with a break in it
8. a circuit with no breaks
9. materials that electric current doesn't flow through
10. safety devices that open circuits to prevent a fire

- **a.** circuit breakers
- **b.** insulators
- **c.** closed circuit
- **d.** circuit
- **e.** open circuit

C50

Understanding What You Learned

1. Plastic covering acts as an insulator to keep what from happening?
2. What is the purpose of circuit diagrams?
3. What is the easiest and most common way to open circuits?
4. How does the current move in a series circuit?
5. How is an electromagnet like a permanent magnet?

Applying What You Learned

1. Why does a lightbulb get hot when it is turned on?
2. Explain how magnets and electricity can be made to work together.
3. Why is it better to have the lights in your home on parallel circuits?
4. How can you create a magnetic field?

5. How does electric current move?

For Your Portfolio

Think about how electrical energy is used in our everyday lives. Give examples of different ways electrical energy is used wherever we go. For example, in a restaurant, the lighted sign outside would use series circuits, the inside lights would use parallel circuits, and electromagnets could help run the motors in the restaurant.

CHAPTER 3

Heat

What comes to mind when you think about heat? Do you picture the bright sun blazing on a hot summer day? Or perhaps you think of holding a hot cup of soup in your hands to warm them on a cold winter afternoon. You might think of a fireplace. All of these scenes show how we use heat.

The most powerful source of energy is the sun. The sun's energy travels millions of kilometers, and only a small part of it reaches Earth. Even so, we can feel the warming effect of sunshine. We know heat moves. Stand near a campfire, and you'll feel the heat. Put a slice of bread in a toaster, and the heat from the electrical coils toasts the bread. Heat and sources of heat can be found everywhere.

The Big IDEA

Heat is produced and transferred in many ways.

CHAPTER SCIENCE INVESTIGATION

Learn about how well different materials stop the transfer of heat. Find out how in your ***Activity Journal.***

C53

Lesson 1

Heat Production

Find Out
- What heat is
- How we use fuel to produce heat
- What fossil fuels are

Vocabulary
fuels
combustion
fossil fuels

The Big QUESTION

How can heat be produced?

When it's cold out, you can warm yourself by standing in the sunshine or moving close to a fire. You won't get rid of a chill standing in the shade or moving away from the fire. What kind of energy is making you warm?

What Is Heat?

Humans have used heat since early humans learned to make fire. They used fire to cook food, provide light, and warm themselves. Today we use many different ways to produce the heat we need. Some of the heat we use is produced by the resistance of electric current running through wires in toasters, stoves, curling irons, and electric space heaters. We also use heat produced by burning materials such as wood and gas. Much of our electrical power comes from burning fuel, such as coal at a power plant, and using the heat as part of the process to produce electricity. Sources of energy that are burned for heat or other kinds of energy are called **fuels.**

Heat is energy that moves from warmer objects to cooler ones. You know that everything is made up of atoms. Some atoms are combined in groups called molecules. These molecules are always moving. Because they have mass and are moving, these molecules have energy. We notice that energy as heat. The faster the molecules move, the more heat there is. The slower the molecules move, the less heat there is.

The water molecules in the left glass are moving faster than those in the right glass. In which glass is the water warmer?

Molecules are always moving—even the molecules in rocks. Have you ever felt a rock on a sunny day? It probably felt very warm. In the shade, however, the rocks were cooler. The molecules in the warm rock were moving faster than those in the shaded rocks. The sun-warmed rock stays warm for a while even after the sun goes down. Can you guess why?

How Do We Measure Heat?

We use temperature as one indicator of heat. Suppose you have two equal-sized bowls of soup. The bowl with more heat will have a higher temperature.

However, different-sized objects made of the same materials do not have the same amount of heat at the same temperature. Because there are more molecules moving in a larger object, the larger object may have more heat, even if it is at the same temperature as a smaller object.

These bowls of soup are at the same temperature, and they have the same amount of heat.

Even though both soups are at the same temperature, the soup in the kettle has more heat because it has more matter. Which soup do you think will cool faster? Why?

Using Fuel to Produce Heat

Through the years, people have found ways to produce heat when and where they want it. Some people burn wood to warm themselves and cook their food. Burning the wood is a chemical reaction called **combustion.** Combustion produces heat and light. During combustion, fuel rapidly combines with oxygen from the air, and burning occurs. The burning produces a gas (usually carbon dioxide), some water, and ash.

People also burn other fuels to produce heat. Coal, natural gas, and oil are among the most common fuels. In some cases, the fuel is burned in a building. For instance, one type of gas furnace warms air that is then distributed throughout your home. In other buildings, the gas is burned to heat water in a boiler. The water absorbs heat from the gas fire. Then, the hot water is sent to radiators throughout the house for warmth.

Coal, natural gas, and oil can also be burned in power plants. The heat produced is used to generate electricity. We use the electricity to give light, heat, and motion. We use electricity to make heat to toast our bread, to melt the point of a glue gun, and to make hot chocolate. We also use the electricity to run televisions, computers, fans, and elevators.

A wood-burning stove uses wood as fuel. During combustion, the wood rapidly combines with oxygen from the air, and burning occurs. Ash can be found in the stove after the wood has burned.

During the process of generating electricity, two-thirds or more of the energy from burning coal or oil is lost as heat to the atmosphere.

Coal can be burned to produce heat.

Using Fossil Fuels

Coal, natural gas, and oil are **fossil fuels.** These fuels were formed by plants that died and decayed millions of years ago. Over that time, the decomposing plants were buried under many layers of soil. Because of the great pressure, they hardened into coal. Some were heated and turned into gas or oil.

Burning fossil fuels provides the electric power that allows us to live the way we do. However, burning fossil fuels has disadvantages. As coal and oil burn, they release harmful chemicals into the air. These chemicals contribute to air pollution throughout the world. Another disadvantage to burning fossil fuels is supply. Fossil fuels took several million years to form. We are using them rapidly, and someday there may not be enough fossil fuels to burn.

Using energy from the sun

Scientists and engineers are finding ways to make combustion of fossil fuels cleaner. This reduces the amount of air pollution formed. They are also finding more efficient ways to burn the fossil fuels. This allows us to produce more heat from the fuels we burn. While scientists work to make burning fossil fuels more efficient, they are also searching for other fuels to burn and better systems for delivering the heat.

In the future, we may be able to turn to the sun for most of our heating needs. Scientists have developed many ways to capture the energy from sunlight. People have been able to create systems that absorb solar energy and produce heat. The system stores the heat for later use. These systems may offer ways for us to use less fossil fuels.

CHECKPOINT

1. What is heat?
2. How do we use fuel to produce heat?
3. What are some examples of fossil fuels?
 Name two ways heat can be produced.

ACTIVITY

Producing Heat

Find Out
Do this activity to learn how electricity can produce heat.

Process Skills
Experimenting
Controlling Variables
Communicating
Predicting

What You Need

desk lamp

25-, 40-, and 60-watt lightbulbs

ruler

two oven mitts

thermometer

watch or clock

Activity Journal

What to Do

1. Set up the desk lamp so the light from the bulb shines on the desktop. The lightbulb should be 15 cm above the desktop.

   Safety! *Be careful when working with the hot lightbulb. Use oven mitts to handle the hot bulb.*

2. With the lamp off, place a thermometer on the desk in the space where the light will shine. Wait three minutes and record the temperature.

3. Place the 25-watt bulb in the lamp. Position the bulb 15 cm above the desk. Turn the lamp on and allow it to shine on the thermometer. After three minutes, record the temperature. Turn off the lamp.

4. Carefully **replace** the 25-watt bulb with the 40-watt bulb. Use the oven mitts to handle the warm bulb. Remeasure so the bulb is 15 cm above the desk. Turn the lamp on and allow it to shine on the thermometer. After three minutes, **record** the temperature. Turn off the lamp.

5. **Predict** what will happen if you replace the 40-watt bulb with a 60-watt bulb. **Record** your prediction. Then, replace the 40-watt bulb, wait three minutes, and **record** the temperature.

CONCLUSIONS

1. Was heat produced by electric current? How do you know?
2. Did the brightness of the bulb affect how much heat was produced? How? What evidence do you have for your answer?
3. Why was it useful to record the temperature before you turned on the 25-watt bulb?

ASKING NEW QUESTIONS

1. What do you think will happen if you repeat the activity using a 100-watt bulb? Why?
2. Will turning the lights off in a room help to keep the room cooler? Why?

SCIENTIFIC METHODS SELF CHECK

✔ Did I **change** only the lightbulb size?

✔ Did I **predict** and **test** what would happen if a 60-watt bulb were in the lamp?

✔ Did I **record** the temperatures of each trial?

Lesson 2

Heat Transfer

Find Out
- How heat is transferred by radiation
- How conduction transfers heat
- How we use convection to transfer heat

Vocabulary

radiation
conduction
convection

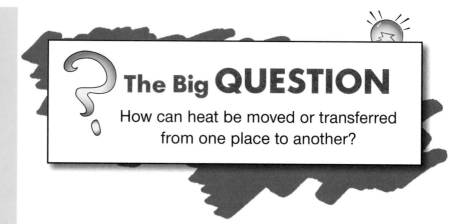

The Big QUESTION

How can heat be moved or transferred from one place to another?

When you sit down to eat a meal, some of the food on your plate may be hot. As you eat, some kinds of food cool down faster than others. Why does this happen? Heat moves away from the hotter food.

If you eat frozen yogurt, you probably don't leave the frozen yogurt sitting around for very long. You know it will melt. But what causes it to melt?

Moving Heat

When you stand near a fire, you can feel the heat without touching the flames. Heat from the fire moves to nearby objects, but not always at the same speed or in the same way. When heat moves from one place to another, it moves from a warmer object to a colder object.

When you stand next to a warm radiator, you feel warmer because heat is moving from the radiator to you. You never get colder while the radiator gets hotter. When you leave your desk to get a book, your chair is warm. Heat has moved from your warm body to the colder chair. When you open the refrigerator and look for a snack, warm air from near the refrigerator door moves into the refrigerator. If you keep the door open long enough, you'll find that the food in the refrigerator has become warm! These three examples show the three ways that heat moves: radiation, conduction, and convection.

Frozen yogurt melts because heat moves from the warmer air to the colder yogurt.

Waves of Heat

Hot objects give off, or radiate, energy in the form of waves. These waves cause the heat to transfer from one place to another. That is why you feel warm when the sun is shining on you. The sun is not touching you, but you can feel heat. The sun's energy moves millions of kilometers through empty space by a process called **radiation** (rā′ dē ā′ shən).

The sun's energy moves to Earth by radiation. The cats can feel the warming effect of the sun.

Moving Atom by Atom

Radiation can transfer heat through empty space. However, heat can be moved in other ways that involve matter. **Conduction** (kon duk′ shən) is the transfer of heat from one group of atoms or molecules to the next group in an object. As the molecules in one spot are heated, they move faster, bumping into other molecules and causing them to move faster. The faster-moving molecules transfer heat through the material. So by heating an object in one spot, the whole object can be heated.

A cooking pan is a good example of heat transfer through conduction. The heat source on a stove touches the bottom of the cooking pan. The molecules from the heat source bump into the molecules from the pan. The pan molecules begin to move faster and the heat spreads through the pan and into the food.

Cooking pans are made of metals that conduct heat well—aluminum, copper, stainless steel, or cast iron. If the pans were not good

The cooking pot is being heated by conduction. As the molecules of the burner move faster, they bump the molecules of the pot, making them move faster. As the faster-moving molecules of the pot bump into the molecules of the soup, the soup heats up. Notice that the aluminum pan is hot, but the wooden spoon is not. Why isn't the spoon hot?

conductors, the food would not cook evenly, because the heat wouldn't travel quickly to all parts of the pan. The food would burn in areas directly over the heat source, and part of the food wouldn't cook at all. Wood and plastic are materials that are not good heat conductors. Their molecules do not transfer heat well.

Moving Air, Moving Heat

Have you ever seen what happens to a curtain hanging above a hot radiator? The curtain moves or flutters. This is because of a convection current. **Convection** (kon vek′ shən) is the transfer of heat by the movement of heated liquids and gases. As hot air molecules move faster and faster, they move apart. This means that hot air is less dense than cold air. As the hot air rises, cold air moves in and takes the place of the rising hot air. The radiator heats the air around it. As this hot air rises, it moves the curtain. When the hot air rises, it is replaced by cold air. The cold air is then heated and rises. The cycle goes on as long as there is a heat source.

The furnace in an older house might use convection currents for heat. The furnace warms the air and the hot air rises to be replaced by colder air. The furnace then heats up the colder air. A modern furnace adds a fan to this system to help move the warmer air throughout the building.

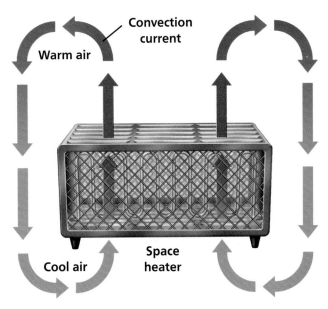

Heated air is moved through the room by convection currents.

Warm air
Convection current
Cool air
Space heater

More Convection Currents

Convection currents also occur in liquids. When a pot of cold water is put on a stove, the heat is conducted through the pan, and the cold water at the bottom heats and begins to rise as it expands. The colder water at the top sinks to the bottom of the pan and is then heated. This creates the movement, or the convection currents, that leads to the heating of all the water.

Convection also explains why your refrigerator will get too warm if the door is left open. When the door is open, the cold air flows out and down and is replaced by lighter, warmer air. As this air is cooled, it also flows out and down. Can you explain why a chest freezer is more efficient than an upright freezer?

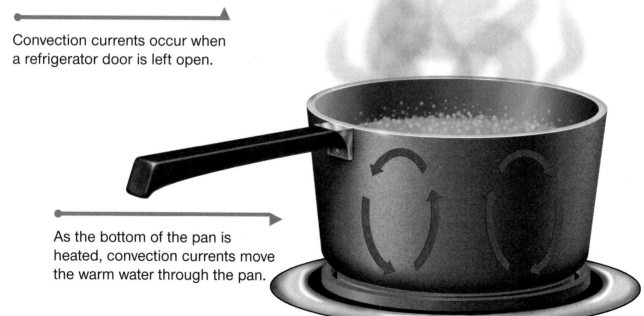

Convection currents occur when a refrigerator door is left open.

As the bottom of the pan is heated, convection currents move the warm water through the pan.

Controlling Heat Transfer

We often want to control how hot or cold things get. Knowing how materials transfer heat can help us do that. Remember that some substances conduct heat better than others. In general, metals are the best heat conductors.

Sometimes a good conductor is not what you want. Most cooking pans have wooden or plastic handles that stay cool so that you do not burn yourself when the pot is hot.

Materials that do not conduct heat well are called insulators. Air, wood, newspaper, and wool are good heat insulators. Many insulators contain trapped air spaces. Convection currents can transfer heat quite well in air, but when the air is trapped in tiny pockets, large convection currents cannot form. Thus, the heat transfer is slowed.

Insulation in buildings uses tiny air pockets to stop the movement of heat. This insulation helps keep buildings warmer in winter and cooler in summer. Insulation is usually placed in the walls and ceilings of homes to reduce the cost of heating and cooling. The insulation also saves energy resources. For many reasons, it's important to conserve valuable energy resources.

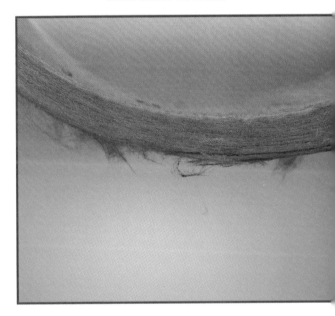

The air pockets trapped in this fiberglass insulation don't allow convection currents to transfer heat. The fiberglass itself is a poor conductor of heat.

CHECKPOINT

1. How is heat transferred by radiation?
2. How does conduction transfer heat?
3. How is heat transferred by convection?
4. How can heat be moved or transferred from one place to another?

ACTIVITY

Observing Heat Transfer by Conduction

Find Out
Do this activity to learn how different types of material affect the way heat is transferred.

Process Skills
Predicting
Communicating
Observing

What You Need

hot water, two butter pats, metal washers, two hand towels, two sturdy pie pans, plastic spoon, 14 cm of aluminum foil, wooden craft sticks, *Activity Journal*

What to Do

1. Make an aluminum "stick" by folding the aluminum foil in half, and then in half again four more times. The wooden stick and the aluminum stick should be about the same size.

2. Place one butter pat on the end of each stick. Put a hand towel under each pie pan.

3. **Predict** which butter pat will melt faster when the sticks are placed in hot water—the butter on the aluminum stick or the butter on the wooden stick. **Write** your predictions.

Safety! *Be careful when handling hot water.*

4. Carefully fill each pie pan with hot water. Place the sticks in, keeping the butter out of the water. Drop washers on the end of each stick under water to help keep the stick in place. Use the plastic spoon to move the washers if you need to.

5. **Observe** what happens to the butter pats. **Record** your observations.

6. Feel the ends of the two sticks. Does one feel warmer than the other? **Record** your observations.

CONCLUSIONS

1. **Compare** your prediction with your observation.
2. What form of heat transfer melted the butter?
3. Why did the sticks need to hold the butter outside of the pie pan?

ASKING NEW QUESTIONS

1. **Predict** what will happen if you put the butter pat on the end of the plastic spoon. Test your prediction.
2. How would the towels feel after you remove the pie pans holding hot water? Why?

> **SCIENTIFIC METHODS SELF CHECK**
>
> ✔ Did I **predict** what would happen when the spoon was in hot water?
>
> ✔ Did I **observe** what happened to the butter?
>
> ✔ Did I **record** my prediction and my observations?

Lesson 3

Using Heat

Find Out
- How steam can move a train
- How heat can be used to generate electricity
- How heat is used to move cars and hot-air balloons

Vocabulary
steam

The Big QUESTION

How can heat make things move?

You have learned that heat is transferred in different ways. Heat can warm your home, your soup, or a car. But heat can also be used to move other objects and to generate electrical energy.

Heat Causes Change

When water gets hot enough, it will boil. When water boils, the molecules at the surface of the water evaporate and become **steam,** which is water vapor heated to above 100 °C. As steam comes in contact with cooler air, it begins to cool. As it cools, it forms tiny droplets of water. This is condensation. We see this condensation as a white cloud.

Because steam is very hot, it can be used in different ways. Like hot water, steam can be piped through a home to warm it. Steam can also be used to make vehicles, such as trains, move.

As the steam gets hot, the molecules begin to move faster and faster. This movement makes the steam expand. If the steam is kept in pipes, there is no room for it to expand, so great pressure builds up. In early steam engines, the hot, expanding steam was piped into one side of a cylinder. As the steam expanded, it pushed on a piston. While the steam expanded, it was cooled by the surrounding metal. The molecules in the cooling steam would start to slow down, so the piston would stop moving forward. Then the cooled steam was piped out while more pressurized steam was let into the other side of the cylinder. The cycle of expanding and condensing steam pushed the piston back and forth and turned the wheels of the steam engine. In this way, steam was used to make the train move.

An early steam engine

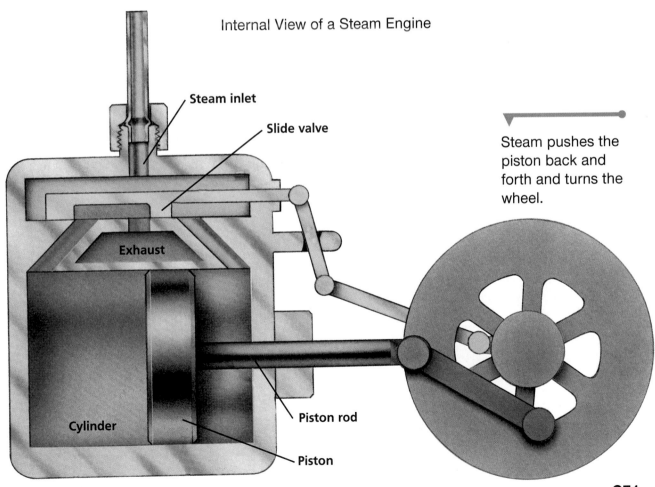

Internal View of a Steam Engine

Steam pushes the piston back and forth and turns the wheel.

C71

Generating Electricity

Steam engines are no longer common but we still use steam in other ways. Many electricity-generating plants burn coal or oil to boil water to make steam. The steam is used to generate another form of energy—electricity. This electricity is the source of the electric current that flows to your home.

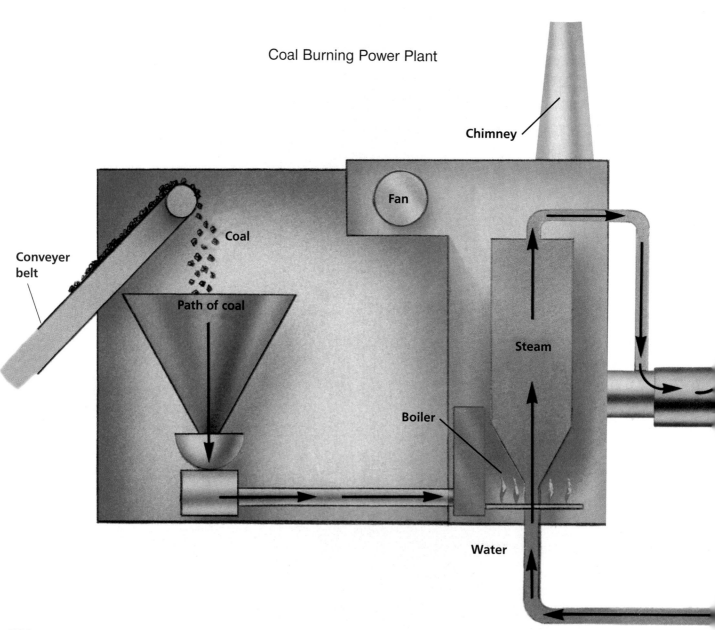

Coal Burning Power Plant

Steam moves the turbines in a power plant just like the girl's breath moves the pinwheel.

In a coal-burning power plant, coal is burned to heat water. As the water molecules start to move faster, they become steam. The hot steam moves around turbines, which are giant fans with many blades. The steam makes the turbines turn just like the wind makes a fan or pinwheel turn. The spinning turbines turn generators to produce electrical energy. As the steam cools, the molecules slow down. The water is piped back to the boiler, where it is heated, and the process starts over again.

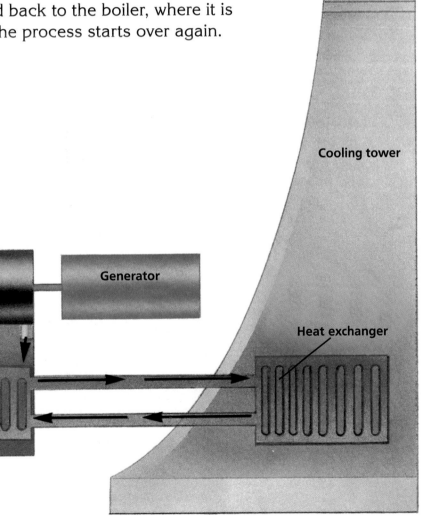

Heat Moves Objects

Moving Cars

The source of motion for a car is similar to the steam used to move old steam engines. A car has a motor with pistons and cylinders. In a car, gasoline burns rapidly inside the cylinders of the engine. The combustion releases hot gases with rapidly moving molecules. These gases expand quickly in the cylinders. The expanding gases push the pistons down. When the hot gases begin to cool, they are forced out of the cylinder. As the gases move out, the pistons move up again. This process is repeated over and over. The up-and-down motion of the piston in the cylinder is turned into the motion that moves the car forward.

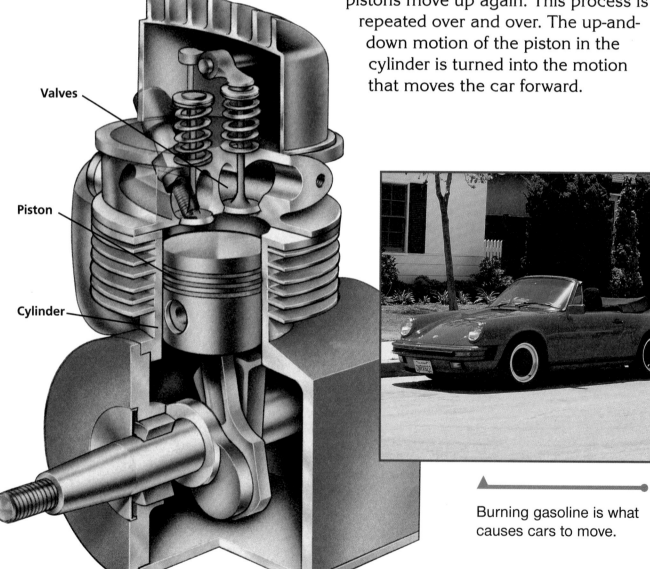

One-Cylinder Gasoline Engine

Valves
Piston
Cylinder

Burning gasoline is what causes cars to move.

Moving Hot-Air Balloons

The fact that gases expand when heated can also explain how a hot-air balloon flies. Remember that as the molecules in the air start to move faster, the air expands and becomes less dense. When the operator of a hot-air balloon turns the burner on, the air in the balloon heats up. After a while, this hot, less-dense air will completely fill the balloon. The balloon will now be less dense than the cool air around it. As a result, the denser cool air pushes up on the balloon and lifts it into the sky.

When the burner is turned off, the molecules begin to slow down, making room for more air molecules to enter the balloon. Eventually the air in the balloon will become just as dense as the outside air, so the balloon will float down. The operator of the balloon can make the balloon rise and fall by heating and cooling the air.

Heating the air in a hot-air balloon makes the balloon rise.

CHECKPOINT

1. How can steam move a train?
2. How can heat generate electricity?
3. How can heat move a hot-air balloon?
 How can heat make things move?

ACTIVITY

Using Heat to Do Work

Find Out

Do this activity to learn how expanding and contracting air can be used to do work.

Process Skills

Observing
Communicating
Predicting
Designing Investigations

WHAT YOU NEED

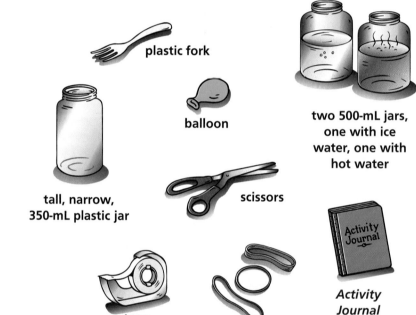

plastic fork

balloon

two 500-mL jars, one with ice water, one with hot water

tall, narrow, 350-mL plastic jar

scissors

tape

rubber bands

Activity Journal

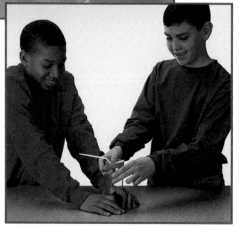

WHAT TO DO

1. Cut off the neck of the balloon. Stretch the balloon tight and flat over the top of the tall jar. Make sure the balloon is centered on the jar top.

2. Put a rubber band around the neck of the jar to hold the balloon in place.

3. Place the fork face up on top of the balloon with the points just over the edge of the jar top. The bowl of the fork should be at about the center of the balloon. Then tape the fork to the edge of the jar so the tape acts like a hinge.

4. Put the jar with the balloon on it as far as possible into a 500-mL jar half full of ice water and hold it there. **Observe** what happens to the end of the fork. **Record** your observations. Take the tall jar out of the water.

5. **Predict** what will happen when the jar with the balloon is moved to hot water. **Record** your prediction.

Safety! *Be careful when handling hot water.*

6. Put the jar with the balloon into a 500-mL jar half full of hot tap water as far as possible and hold it there. **Observe** what happens to the end of the fork. **Record** your observations.

CONCLUSIONS

1. What happened to the air in the jar when you chilled it?
2. What happened to the air in the jar when it warmed?
3. Describe what happened to the handle of the fork.

ASKING NEW QUESTIONS

1. How could you use your jar to lift objects?
2. Think about what you know about water heating on the stove. Would the jar and fork work as well if the jar were completely full of water instead of air?
3. What would happen if you actually boiled water inside the jar?

SCIENTIFIC METHODS SELF CHECK

✔ Did I **observe** and **record** what happened?

✔ Did I **predict** what would happen when the jar with the balloon on it was placed in the hot water?

✔ Did I **design** a way to use the jar to lift objects?

Review

Reviewing Vocabulary and Concepts

Write the letter of the answer that completes each sentence.

1. Sources of energy that are burned for heat or energy are called ___.
   - a. oxygen
   - b. fuels
   - c. fire
   - d. solids

2. During ___, fuel combines with oxygen from the air, and burning occurs.
   - a. convection
   - b. conduction
   - c. radiation
   - d. combustion

3. Heat transferred from the sun to Earth occurs through ___.
   - a. nuclear energy
   - b. radiation
   - c. conduction
   - d. electricity

4. ___ is the transfer of heat from one group of atoms or molecules to another group.
   - a. Conduction
   - b. Locomotion
   - c. Action
   - d. Commotion

Match each definition on the left with the correct term.

5. water vapor heated to above 100 °C
6. coal, natural gas, and oil
7. heat transfer through movement of heated liquids and gas
8. materials that do not conduct heat well

- a. convection
- b. steam
- c. fossil fuels
- d. heat insulators

C78

Review

Understanding What You Learned

1. How does the sun's energy travel through space?
2. Which material is a better heat conductor, metal or plastic?
3. When heat is transferred through convection, what happens to the hot air? The cold air?
4. How can heat be used to generate electricity?

Applying What You Learned

1. Where do fossil fuels come from?
2. How does insulation control heat transfer?
3. Describe the ways that heat is produced and transferred.

For Your Portfolio

Think about ways in which you use energy. Are there ways that you can be more efficient in your use of energy? Write down five things you could do to conserve energy in your daily life.

CHAPTER 4

Matter, Motion, and Machines

Matter is anything that takes up space and has mass. Some examples of matter are air, bicycles, apples, oil, and rocks. In short, matter is all around us.

Matter can be observed, described, and measured. It can also be moved, which involves work. Using simple machines can make work easier to perform.

The Big IDEA

Because matter has mass, it takes force to move it.

CHAPTER SCIENCE INVESTIGATION

Learn how speed and acceleration are related. Find out how in your *Activity Journal*.

LESSON 1

Matter

Find Out
- What matter is
- What volume is
- How mass and weight are different
- What density is

Vocabulary
matter
volume
mass
weight
density
buoyancy

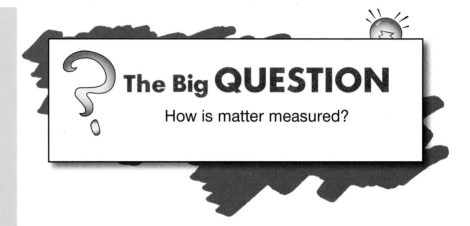

The Big QUESTION
How is matter measured?

*H*umans are constantly changing their ideas as they learn new things. For example, people used to think everything was made of four elements—air, fire, earth, and water. Now we know that everything is actually made of atoms and molecules. Atoms and molecules join in different ways to make different forms of matter.

Describing Matter

You know that matter is all around us, but how can we describe matter? Imagine that you have asked for a new jacket. You could describe the jacket by some of its properties—size 8, red or blue. These properties relate to size and color. Other properties can also be used to describe matter.

One property of matter is that it takes up space. A balloon takes up space, so it must be made of matter. However, the air inside the balloon also takes up space. When a balloon is blown up, the air inside makes the balloon expand. The air takes up space so it is also matter. **Matter** is anything that takes up space and has mass.

Volume

Another way to describe matter is to look at its size. You can measure the size of something by how much space it occupies. **Volume** is the amount of space something takes up. Volume is often calculated in cubic centimeters. You can measure the volume of a book, a box, an aquarium, or a balloon. To find the volume of a box, you need to measure its height, width, and length. Then, multiply these three numbers together to find out how much space the box takes up.

Scientists also use other methods and tools to measure the volume of liquids or odd-shaped solids. A beaker can be used to find the volume of a liquid. After pouring the liquid in, you can look at the numbers on the side of the beaker to find the volume. A graduated cylinder is another type of container used to measure volume. If you drop a solid into a liquid in the cylinder, you will notice a change in the level of the liquid. By subtracting these two numbers, you can calculate the volume of the solid.

Both of these containers have 300 mL of water in them. Can you calculate the volume of the rock?

This box is 24 cm high, 21 cm wide, and 30 cm long. Its volume is 24 cm x 21 cm x 30 cm = 15,120 cubic centimeters.

Moon

180 kg

300 N

1800 N

Mass and Weight

How can you describe a golf ball and a table-tennis ball? Look at the photo below. The golf ball and the table-tennis ball are nearly the same size, so they would have about the same volume. They are also the same color. However, if you hold the two balls in your hands, you will notice that they feel very

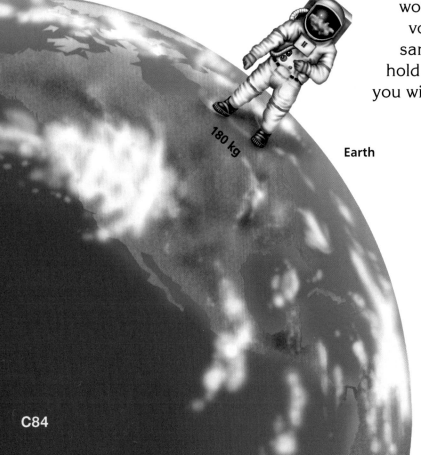

180 kg

Earth

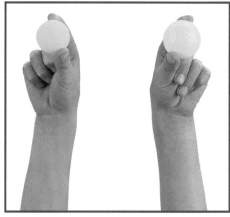

different. One reason is that the golf ball has more matter than the table-tennis ball. The amount of matter in an object is the **mass** of the object. Because the golf ball has more matter, we say that it has more mass. Mass is a property of matter. It is measured in grams or kilograms.

When describing matter, you might also hear people refer to weight. Weight is another property of matter that can be measured. Weight and mass are often confused, but they are not the same. **Weight** is the amount of force that gravity exerts on an object and is measured in newtons. For example, you weigh more than a golf ball. That means that Earth's gravity is pulling downward on you more than it does on a golf ball. Because the golf ball has more mass, it also weighs more than the table-tennis ball. However, objects that have the same mass can also have different weights. Look at the illustration of the astronaut on Earth, on the moon, and on Mars. You can see that the mass of the astronaut does not change from place to place. However, because gravity at each place is different, the astronaut's weight does change.

684 N

Mars
180 kg

Mass is measured in kilograms (kg). Weight is measured in newtons (N). The mass of the astronaut stays the same. What happens to the astronaut's weight on Earth, the moon, and Mars?

Density and Buoyancy

Water is denser than oil.

Some matter takes up a large space but has very little mass. A beach ball can have a large volume, but the ball and the air inside can have a very small mass. You can easily pick up a beach ball because of the small mass. Other objects can take up very little space but have a large mass. A bowling ball can have a smaller volume than a beach ball, but it has much more mass.

When you compare the amount of matter with the space that it takes up, you are looking at another property of matter. **Density** (den′ si tē) is the measure of the mass of a material for a specific volume. Gases usually have a very low density. A bowling ball has a much higher density than a beach ball.

One easy way to compare the densities of objects is to put them in water. If the object sinks, it has a greater density than water. Materials that are less dense than water will float. Look at the photo of the measuring cup with water and oil. If we compare the corn oil and the water, we can say that the water has a density of 1. The corn oil has a relative density of 0.9. Since the corn oil is less dense, it floats on the water.

Solid objects can also be less dense than water. Most human bodies are a little less

Which is denser than water—a table-tennis ball or a golf ball?

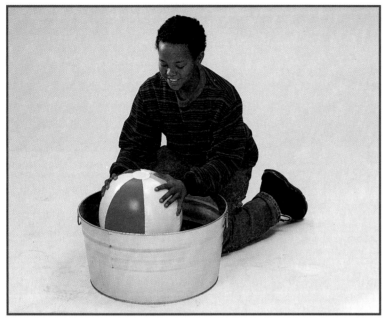

Beach balls are very buoyant.

dense than water. That is why people can float in water. If you put your face under water and blow out air, the bubbles will also float to the top of the water. Air is less dense than water so the bubbles rise in the water.

The ability of an object to float or rise in a liquid is called **buoyancy** (boi′ ən sē). When you blow bubbles under the water, the force of the water pushes up against the weight of the bubble. If you think about the beach ball again, you can feel the effect of buoyancy. Place the beach ball in water. Now push down on it. You can feel the force of the water pushing up. A way to increase the buoyancy of something is to add air to it. Since gases are not very dense, they float in water easily. A beach ball contains a lot of air so it is very buoyant.

CHECKPOINT

1. What is matter?
2. What is volume?
3. Explain the difference between an object's mass and its weight.
4. What is density?

 How is matter measured?

C87

ACTIVITY

Comparing Density

Find Out

Do this activity to see how volume and density of an object can affect an object's buoyancy.

Process Skills

Measuring
Communicating
Inferring
Predicting
Designing Investigations

What You Need

pan balance

stack of pennies

aluminum foil

metric ruler

container or bucket

water

What to Do

1. Measure and cut two pieces of aluminum foil to 10 cm by 10 cm.
2. Place a penny on each sheet of aluminum foil.
3. Wad one sheet of foil tightly around the penny.
4. Use the other sheet of foil to make a boat for the penny to float in water.

5. **Estimate** and then **calculate** the mass of each penny-and-foil pair. **Record** your findings.
6. Gently place the foil boat with the penny in a bucket of water. **Record** what happens.
7. Now gently place the wad of foil with the penny in the same water. **Record** what happens.

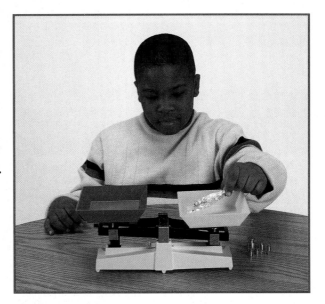

CONCLUSIONS

1. Which foil-and-penny pair had a greater mass?
2. Which pair had a greater density than water? How do you know?
3. Why do you think the two pairs acted differently in water?

ASKING NEW QUESTIONS

1. If the mass of an object stays the same but the volume decreases, what happens to the density?
2. **Predict** what shape of aluminum boat will hold the most pennies in water. **Design an investigation** to test your prediction.

SCIENTIFIC METHODS SELF CHECK
- ✔ Did I **estimate** and **measure** the mass of the aluminum foil-and-penny pairs?
- ✔ Did I **compare** how the two pairs acted in water?
- ✔ Did I **record** my measurements and observations?
- ✔ Did I **test** my predictions?

LESSON 2

Motion

Find Out
- What inertia is
- What friction does
- How force affects the motion of an object

Vocabulary
force
inertia
friction
speed
acceleration

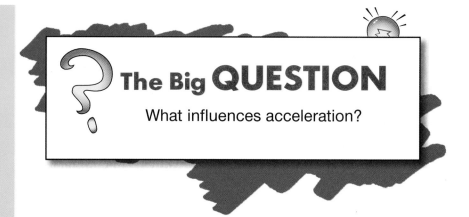

The Big QUESTION
What influences acceleration?

*N*ow you know that all objects are made of matter, have mass, and occupy space. But how can an object start to move? Once an object is moving, how can it be stopped? Forces must act on the object.

Inertia

Have you ever started to run down a hill only to find that you can't stop yourself? Your body continues to go forward even though you want to stop. The reason this happens is the same reason that the rider is falling off the horse in the photo. The horse and the rider both have mass. They were both moving forward, but the horse stopped. Nothing stopped the rider, however, so he will continue moving forward until a force causes him to stop. This shows a basic property of all matter. Moving matter tends to keep moving until it is acted on by another force.

A **force** is a push or a pull. The horse gave the force that pushed the rider to start moving. Gravity is another force that always pulls the rider downward. The rider's motion will not stop until the force of the ground pushes against his body. Matter needs a force to stop it from moving. But how can you get matter to move when it isn't moving? Again, the answer is force.

Matter tends to keep doing what it was already doing. It takes force to change that. Imagine that a cotton ball is sitting on your desk. The cotton ball will not move unless you push it (or the wind blows it or some other force pushes or pulls it). Now imagine a bowling ball sitting on your desk. It will not move until a force pushes or pulls on it. But the force needed to move the bowling ball is much greater than the force needed to move the cotton ball. Why? The mass of the bowling ball is much greater than the mass of the cotton ball. An object with more mass has more inertia. **Inertia** (ə nûr′ shə) is the tendency for matter that is not moving to stay not moving, or if it is already moving, to stay moving.

Inertia causes the rider to continue moving even though the horse has stopped.

Friction

Friction between the tires and the ground stops the bike from skidding and moves the bike forward when you pedal.

Moving things also speed up, slow down, and change directions. Think about riding your bike. Your bike sits in the garage until some force moves it. When you push down on the pedals, you are using your legs as a force. But why doesn't the bike just continue moving after the first push? Another force that is pushing on the bike is friction. **Friction** (frik′ shən) is the resistance force created between the surface of a moving object and its surroundings. Friction is the force that stops a ball that is rolling on the ground. Friction will stop your bike if you don't pedal.

Imagine that as you pedal, you come to a hill. Why do you have to pedal harder to go up the hill? You are already pushing against the force of friction, but now you also have to push against the force of gravity. As you go up the hill, the force of gravity is pulling against both you and your bike, and friction is creating resistance. Your legs have to work harder to overcome both forces. That is why it is harder to pedal your bike up a hill than it is to pedal your bike on a flat street.

Snowboarders often try to reduce the amount of friction so that they can travel at higher speeds on the snow.

Friction also helps to stop your bike from skidding down the hill. The resistance helps to hold the tires on the pavement. Water, grease, and ice all reduce the amount of friction. Imagine riding your bike on the ice. Without enough friction, it would be impossible to ride up an icy hill, no matter how hard you pedal.

Force and Motion

Speed measures how quickly an object moves. Or, to put it another way, speed measures how far an object moves during a certain time. For example, a truck might travel 50 km in an hour.

Speed is also related to force. Look at the baseball player. He can tap the ball gently or hit it as hard as he can. The force of the hit will determine the speed of the ball.

A force can speed up an object, slow down an object, or change its direction. In baseball and softball, the bat is used to change the speed of a pitched ball. The catcher's mitt, on the other hand, is intended to stop the moving ball. The catcher's mitt is well padded so it can absorb the force of the moving ball. When a ball is moving at a high speed, the force needed to stop it can be very painful to your bare hand.

The force of the swinging bat can change the speed of the ball. The catcher's mitt absorbs the force of the moving ball.

Moving faster or slower also involves force and mass. If a soccer player kicks a ball, then the ball will accelerate. **Acceleration** (ak sel′ ər ā′ shən) is the rate at which the speed of an object changes. A kick from the player can accelerate the ball by slowing it down, speeding it up, or changing its direction. A hard kick will increase the acceleration. A soft kick causes a smaller acceleration.

What if we use a ball that has a different mass? Acceleration depends on the force acting on an object and the mass of the object. If you replace the soccer ball with a bowling ball, a soft kick may not move the ball. More force is needed to move an object with more mass. We expect matter to move in predictable ways, so we plan our actions to fit our predictions. You would kick a soccer ball, but not a bowling ball. You know that the force of your foot is not strong enough to move the mass of the bowling ball, and if you tried, you might get hurt.

A hard kick will increase the acceleration of the ball.

A soft kick will cause a smaller acceleration.

Engineers think about the relationships between mass, force, and acceleration when they design devices. For example, road bicycles are built lightweight because they travel mostly on smooth roads. Their light weight means they take less force to accelerate. Mountain bikes, on the other hand, have more mass so they can withstand the impact of riding over rocks and bumpy ground. This makes mountain bikes heavier. So it takes more force to accelerate a mountain bike.

The relationship between mass, force, and acceleration can also be observed in nature. Have you ever tried to catch a grasshopper? Grasshoppers are hard to catch because of their great ability to accelerate. They have a small amount of mass, so they only need a small force to accelerate. Their small hind legs and wings produce enough force to accelerate them quickly. Having a low mass allows grasshoppers to change directions almost instantly. Grasshoppers demonstrate that races are not always won by the fastest competitor. Sometimes the best accelerator wins.

It takes more force to accelerate the man's bike because he and his bike have more mass than the girl and her bike do.

Grasshoppers can accelerate very quickly.

CHECKPOINT

1. What is inertia?
2. What does friction do?
3. Describe how a force affects the motion of an object.
    What influences acceleration?

ACTIVITY

Observing Inertia

Find Out
Do this activity to see that moving objects have inertia.

Process Skills
Constructing Models
Observing
Communicating
Measuring
Inferring
Predicting

What You Need

small piece of modeling clay

small toy car that can roll on the ruler

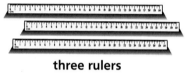
three rulers

masking tape

pencil

two or three thick books

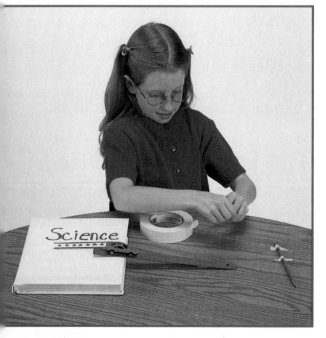

What to Do

1. Place the ruler on a desk. Use a book to prop up one end of the ruler.
2. Tape the other end of the ruler to the desk.
3. Place the pencil perpendicular to the ruler. Move it about two car lengths away from the ruler. Tape the pencil at each end to the desk.
4. Make a small clay figure out of the modeling clay. Put the clay figure on the hood of the toy car. Do not press the clay into the car. Place it gently on the hood.

5. Put the car with the clay figure on the high end of the ruler.
6. Let go of the car. **Watch** as it rolls down the ruler and collides into the taped pencil. **Record** your observations.
7. Use the second ruler to **measure** how far the clay figure falls from the car.
8. Repeat steps 5–7 several times. **Record** your results each time.
9. Use the second book to prop the ruler up higher. Repeat steps 5–7 several times. **Record** your results each time.

CONCLUSIONS

1. What happened to the clay figure and the car when they collided with the ruler?
2. What did you **observe** when you raised the ruler higher?
3. Why do you think this happened?

ASKING NEW QUESTIONS

1. **Predict** what will happen if you place the pencil farther away from the end of the ruler. Test your prediction.
2. Why should people wear seat belts in cars?

SCIENTIFIC METHODS SELF CHECK

✔ Did I **record** my observations each time?

✔ Did I **repeat** the activity several times before raising the ruler?

✔ Did I **measure** how far the clay figure moved?

✔ Did I **test** my prediction?

Simple Machines

Find Out
- What simple machines have in common
- How levers and wheel and axles are useful
- How pulleys, inclined planes, wedges, and screws are useful

Vocabulary
machine
lever
wheel and axle
pulley
inclined plane
wedge
screw

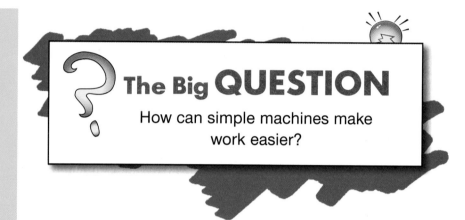

The Big QUESTION

How can simple machines make work easier?

*W*hat machines have you used today? Probably many more than you think. Every day, you use machines as you play and work. Machines can help you change or increase forces to make work easier. They help us do things we can't do on our own.

Six Simple Machines

A **machine** is a device that performs work. An automobile is a very complicated machine, with more than 15,000 different parts. However, a machine doesn't need to be complicated to be useful. There are many simple machines that you see and use every day. They include the seesaw at the playground, the pedals and sprocket on your bicycle, the pencil sharpener in your classroom, the scissors used to cut your hair, and the screws used to hold your chair together.

Although there are many examples of simple machines, they all have several things in common. They have few moving parts. They do not need electricity to work. Instead, people make them work. Simple machines can be used together with other machines, or they can be used alone.

Simple machines make work easier. Work is done when a force acts on an object and moves it over a distance. Simple machines don't do more work than people do to use them. Simple machines just make the work easier.

There are six commonly used simple machines: lever, wheel and axle, pulley, inclined plane, wedge, and screw. You come in contact with all of these machines every day, but you may not notice them. A simple machine can reduce the force needed to do a job. It can change the direction of the force to make work more convenient. With other simple machines, you can move one part of the machine slowly as another part of the machine moves quickly.

A piano uses levers to strike the strings, which make the sound.

A pull-top can is a lever.

Levers and Wheel and Axles

Levers

A **lever** is an arm that turns around a point. Tools that pry or pull things up are levers. It is easier to lift a very heavy object or pry something up with a lever than it is without one. A screwdriver can pry the lid off a paint can and a hammer claw can pull out a nail. When you open a can with a tab, you use a lever. Your finger pulls up the tab, which pushes down and opens the can. The tab is the lever that turns on the top of the can when you lift it.

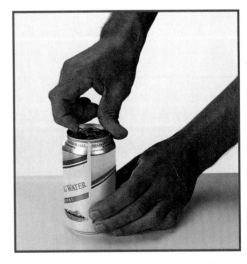

A wrench is a lever. The longer handle can give you more force to turn the nut.

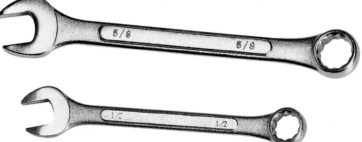

The point, or pivot, the lever turns on is called the fulcrum (fool′ krəm). The base of the seesaw at the playground is a fulcrum. The fulcrum doesn't move, but the lever moves up and down when force is applied.

A wrench is also a lever. When you turn a nut on your bike wheel with a wrench, you use a smaller force on a long handle to get a larger force to turn the nut. If you need more force to to turn the nut, you can get a wrench with a longer handle. This handle gives you more "leverage."

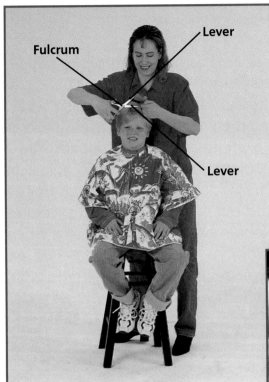

Each blade acts like a lever. The sharpened edges of the blades can cut material or hair.

The scissors used to cut your hair are also levers. Each blade is a separate lever. They are attached together and hinged at the fulcrum. By using one hand, you apply force on both levers at the same time.

By applying force to one end of a lever, you can gain speed and distance at the other end. When you use a hammer to drive in a nail, you are holding close to one end so that you can move the hammer head upward quickly and let it crash down rapidly on the nail head. A fishing rod is used in the same way. You use one hand as the fulcrum on one end of the rod to get the tip of the rod to move rapidly to cast the line.

By holding one end of the fishing rod, the other end can gain speed. This lever can cast the fishing line and can help you lift the weight of the fish.

Wheel and Axle

A **wheel and axle** is a simple machine with a wheel that turns a post. The post is called the axle. A wheel and axle is really just a lever that can turn all the way around. Turning the wheel produces a greater force on the axle. Many common objects use a wheel and axle. For example, a doorknob pulls the latch open with a wheel and axle. The pedals and sprocket that move the chain on your bicycle are a wheel and axle as well. So is the steering wheel on a car or the faucet on a sink. Almost anything with a crank attached works as a wheel and axle.

A doorknob is a wheel and axle. When you turn the knob, it pulls the latch over so the door can be opened.

Turning the handle exerts the force that sharpens your pencil.

Pulleys, Inclined Planes, Wedges, and Screws

Pulley

With a **pulley,** a rope is placed over a grooved wheel. Most often a pulley is used to lift things. In construction projects, for example, a system of pulleys and ropes can be used to lift or lower heavy objects, such as a piano, up or down the side of a building.

A single pulley makes work easier by allowing you to use your weight to lift an object connected to the other end of the rope. The pulley doesn't increase the force. It just allows the pulling to be done in an easier direction. A double pulley uses the same rope to wrap around two grooved wheels. The force lifting the load is doubled with the extra pulley, so you only have to use half the effort to lift the same load.

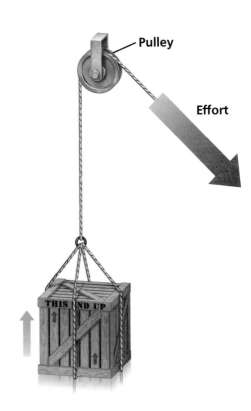

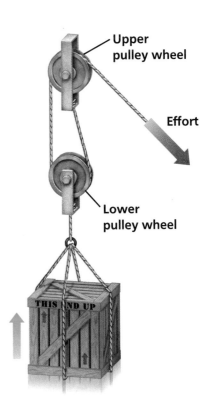

Pulleys make work easier.

It is easier to push a load up a ramp than it is to lift it.

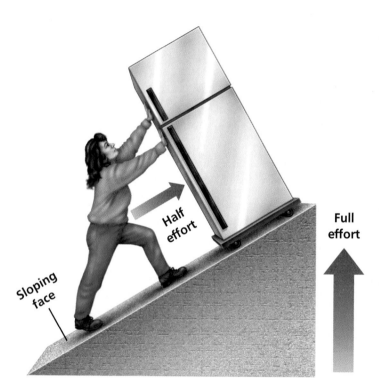

Inclined Plane

An **inclined plane** (in′ klīnd plān) is a flat surface that is higher at one end. *Inclined* means "at a slant," and *plane* means "flat surface." Using an inclined plane makes it easier to slide a load upward than to lift it directly. Ramps, which are often used alongside stairways for wheelchairs or baby strollers, are inclined planes. It doesn't take as much force to move a wheelchair or stroller up a ramp as it does to lift it up step by step. When a ramp is twice as long as it is high, the force needed to move an object to the top is half of the force that would be needed to lift it up.

Wedge and Screw

An inclined plane used to push things apart is called a **wedge** (wej). A doorstop works as a wedge. So does a knife used to slice cheese or cut bread, because the knife pushes the cheese or bread apart. An ax is a wedge that separates a log into two pieces.

Another type of inclined plane is a **screw.** A screw is an inclined plane that winds around a rod into a spiral. A screw holds things together more strongly than a nail does. To come apart, the objects must either break in two or move the whole length of the screw threads as the screw is turned.

Screws are usually used to hold things together. When you tighten the lid on a jar, for example, you are using a screw to pull the lid down against the mouth of the jar.

The lid of the peanut butter jar is a screw.

CHECKPOINT

1. What do simple machines all have in common?
2. How are levers and wheel and axles useful?
3. How are pulleys, inclined planes, wedges, and screws useful?

 How can simple machines make work easier?

ACTIVITY

Using Simple Machines

Find Out
Do this activity to learn how a simple machine can move an object.

Process Skills
Experimenting
Observing
Interpreting Data
Constructing Models
Designing Investigations

What You Need

scissors

masking tape

safety goggles

Activity Journal

building materials

What to Do

1. Find a way to get the roll of tape from the floor to your desktop. You can use your hands to provide the lift. But you can't just pick up the tape and put it on the desk.

2. With your group, think of as many different ways as you can to lift the tape. Then, plan two different ways to get the tape to the top of the desk. **Write or draw** your ideas.

3. Show your plans to your teacher. After your teacher has approved your plans, **try** to construct one of them.

   Safety! *Wear safety goggles if you are using rubber bands.*

4. **Record** what happens. If your plan didn't work well, what went wrong? What could you do to fix it?

5. **Try** your other plan for lifting the tape. **Write** down what happens this time.

6. Discuss your other ideas with your teacher. Then try each of them. **Record** your results each time.

CONCLUSIONS

1. How many ways did you come up with to raise the tape to the top of your desk? Which one worked the best?
2. What materials did you use?
3. What forces did you use?

ASKING NEW QUESTIONS

1. Why didn't some of your ideas work as well as others?
2. Which idea required the least amount of effort to do?

SCIENTIFIC METHODS SELF CHECK

- ✔ Did I **plan models** of different ways to lift the tape?
- ✔ Did I **observe** what happened with each plan?
- ✔ Did I **interpret the data** to see what could be done to make a plan work?
- ✔ Did I **experiment** with different materials to see if I could make the plan work?

Review

Reviewing Vocabulary and Concepts

Write the letter of the answer that completes each sentence.

1. ___ has mass, takes up space, and has properties by which it can be described.
   - **a.** Matter
   - **b.** Inertia
   - **c.** Acceleration
   - **d.** Volume

2. The measure of the pull of gravity on an object is called ___.
   - **a.** weight
   - **b.** mass
   - **c.** density
   - **d.** inertia

3. A measure of how much mass an object has for its size is called ___.
   - **a.** volume
   - **b.** inertia
   - **c.** mass
   - **d.** density

4. The tendency of matter to continue moving or to remain still is called ___.
   - **a.** speed
   - **b.** force
   - **c.** inertia
   - **d.** mass

5. ___ is the resistance created between the surface of a moving object and its surroundings.
   - **a.** Inertia
   - **b.** Friction
   - **c.** Acceleration
   - **d.** Speed

6. The rate at which the speed of an object changes is called ___.
   - **a.** acceleration
   - **b.** speed
   - **c.** inertia
   - **d.** friction

7. A simple machine with a rope fitted around a grooved wheel is a ___.
   - **a.** wheel and axle
   - **b.** lever
   - **c.** pulley
   - **d.** wedge

8. A ___ is an inclined plane used to push things apart.
   - **a.** lever
   - **b.** screw
   - **c.** wedge
   - **d.** pulley

Review

Match each definition on the left with the correct term.

9. the amount of space something takes up
10. the amount of matter in an object
11. the push or pull that one object has on another object
12. how quickly an object moves over a certain distance
13. a device that performs work
14. an inclined plane that winds around a rod into a spiral

a. machine
b. mass
c. force
d. speed
e. screw
f. volume

Understanding What You Learned

1. How does a wedge work?
2. If you open a can with a tab, which simple machine are you using?
3. How could you feel buoyancy using a beach ball?

Applying What You Learned

1. Describe how a skier or snowboarder uses friction to increase or decrease speed or to stop.
2. Would an astronaut's weight and mass change if she traveled from Earth to Mars? Explain.

3. Explain how the motion of matter is related to force.

For Your Portfolio

You have learned how matter moves and how machines work. Choose one of the following titles and write a short story about how matter moved or how a machine performed work: *A Day at the Bike Shop; Our Neighbors Build Their House; Track and Field Events at the Olympics.*

UNIT C

Unit Review

Concept Review

1. How can electricity and magnets make things move without touching them?
2. How are series circuits and parallel circuits similar?
3. Make a list of all the ways that heat is produced and transferred in your own home.
4. You have learned that matter is anything that takes up space and has mass. How can you make matter move?

Problem Solving

1. Suppose you have a magnet that does not have its poles marked. How would you find out which end of the magnet is the north-seeking end?
2. Suppose that you have wrapped wire around an iron object to create an electromagnet. However, the electromagnet doesn't work. What would you check?
3. You know that heat is transferred in three ways: radiation, conduction, and convection. Define each method and explain two different methods you could use to cook an egg.
4. Give an example of each of the six simple machines that you use every day. Explain how you would do this work without the simple machine.

Something To Do

Develop a class presentation to explain the material covered in this unit to children in one of the lower grades at your school. Divide the class into four groups. Each group will be responsible for explaining the important concepts that are presented in one of the chapters. You might use charts and posters, scrapbooks, collections of items from home, demonstrations, role playing, skits, songs, invited speakers, or videotapes.

Practice your presentation in front of your class before doing it for a lower grade level.

UNIT D

Health Science

Chapter 1 **Skin** **D2**
    Lesson 1: **Properties of Skin** D4
      Activity: The Sense of Touch D10
    Lesson 2: **Skin Controls Body Temperature** D12
      Activity: Facial Wrinkles D18
    Chapter Review **D20**

Chapter 2 **Chemical Substances** **D22**
    Lesson 1: **Chemical Substances Cause Changes** D24
      Activity: Educating Others D30
    Lesson 2: **Positive Effects of Chemical Substances** D32
      Activity: Reading the Label D38
    Lesson 3: **Abuse of Chemical Substances** D40
      Activity: Tracing the Effects of Chemical Substances D46
    Chapter Review **D48**

Chapter 3: **Nutrition** **D50**
    Lesson 1: **Carbohydrates, Fats, and Proteins,** D52
      Activity: Finding Fats and Starch in Foods D58
    Lesson 2: **Water, Vitamins, and Minerals** .. D60
      Activity: Reading a Food Label D66
    Chapter Review **D68**

Unit Review **D70**

CHAPTER 1

Skin

Look at the people around you. See how each person is dressed a bit differently from everyone else. Beneath a person's clothing, however, everyone is wearing something that's the same. That "something" is the body's largest organ—the skin!

Even high-quality clothing eventually wears out. Although skin wrinkles, stretches, and scars as it ages, it is meant to last a lifetime. It covers your body and helps to control body temperature. It removes waste and even protects your body from germs. Skin also gives you the sense of touch. Skin is truly an amazing organ!

The Big IDEA

Skin protects the body and controls the body's temperature.

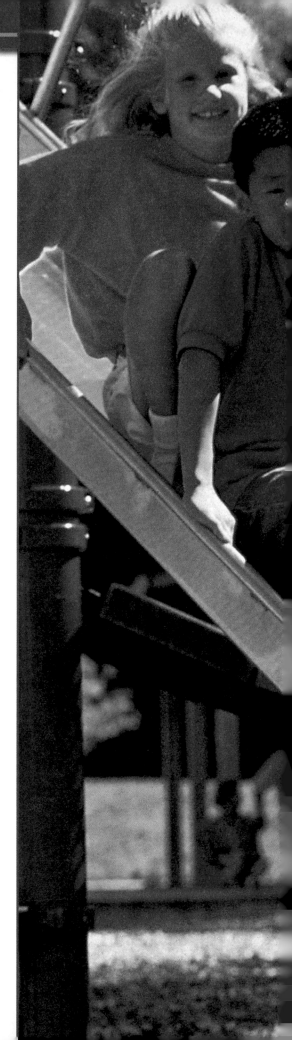

CHAPTER SCIENCE INVESTIGATION

Make a model of your skin and its various parts. Find out how in your *Activity Journal.*

Lesson 1

Properties of Skin

Find Out
- How the layers of your skin differ
- How skin protects your body
- What other work your skin does

Vocabulary
integumentary system
epidermis
melanin
dermis
pores
hair follicles

The Big QUESTION
Why do we need skin?

*I*magine what life would be like if you had to put on a different suit of skin each day. You would wear waterproof skin on rainy days, furry skin on cold days, and lightweight skin on hot days. Believe it or not, your skin already provides for all of these situations—and you don't even have to change it every day.

The Layers of Your Skin

Skin, along with your hair and nails, makes up your body's **integumentary** (in teg yə ment′ rē) **system.** Just like your bones and muscles, the integumentary system protects your body and helps give your body its shape.

And, like everything else in your body, your skin is made of cells. In fact, your skin is constantly shedding cells and replacing them with new ones. You don't have to change your skin—it is constantly changing and growing along with the rest of your body.

The skin has two main layers, the epidermis and the dermis. The **epidermis** (ep ə der′ məs) is the thin outer layer of skin. The top layer of the epidermis is made up of both living and dead cells. The dead cells are constantly being replaced by new cells. There are no blood vessels or nerves in the epidermis. Only cuts that are deeper than the epidermis will bleed.

Skin cells in the epidermis are stacked on top of each other in layers. New skin cells are made in the bottom layer and are slowly pushed to the skin's surface. By the time the new cells reach the surface, they are dry and flat. The body sheds these flakes of dry, dead skin.

Skin cells at the base of the epidermis contain a substance called melanin (mel′ ə nən). The amount of **melanin** in your skin determines your skin color. People with darker skin have more melanin than people with lighter skin do.

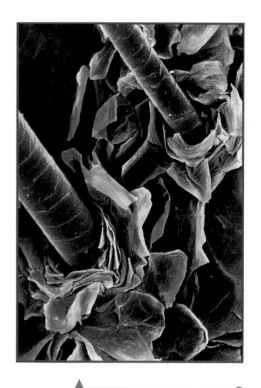

This image shows magnified scalp hairs and the flakes of dead skin on the epidermis.

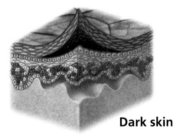
Dark skin

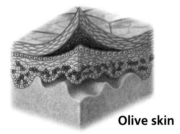
Olive skin

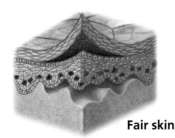
Fair skin

The amount of melanin in your skin determines your skin color.

When you expose your skin to sunlight, the amount of melanin made by skin cells increases. This is how the skin tries to protect itself against the sun's ultraviolet rays. Small increases in melanin can result in a suntan. However, sunlight can also damage your skin. Getting too much sunlight can increase your chance of getting skin cancer. It can also give you a bad sunburn.

The **dermis** (dur′ məs) is the thicker layer of skin under the epidermis. In some places on your body, the dermis is a very thick layer. In other places, the dermis is thinner. The thickest layer of dermis is on the soles of your feet. The thinnest layer of dermis is in your eyelids.

The dermis contains sweat glands, blood vessels, nerve endings, and also the roots of hair. Sweat glands release sweat through openings in the epidermis called **pores.**

Oil glands in the dermis release an oily substance. This body oil keeps your skin and hair from becoming dry. Oil helps waterproof your skin and keep the inside of your body moist. The oil also kills some bacteria that can harm your body.

Hair roots in the dermis grow up through thin tubes called **hair follicles.** Your hair, just like your fingernails and toenails, is dead and is pushed up through the skin's surface. An average hair on your scalp will last about three years before it falls out and is replaced by a new one.

The skin is made of an outer layer—the epidermis—and an inner layer—the dermis.

Your Skin Protects Your Body

Your skin helps to keep out germs and other harmful materials. The oil produced in glands helps to kill some bacteria. The waterproof coating of your skin stops many germs from getting to your other organs. Even your hair stops things from harming your organs. For example, your eyelashes protect your eyes. Your skin does not do a perfect job, however. Some harmful things, such as chemicals, can be absorbed through the skin. But your skin does protect your body in many ways.

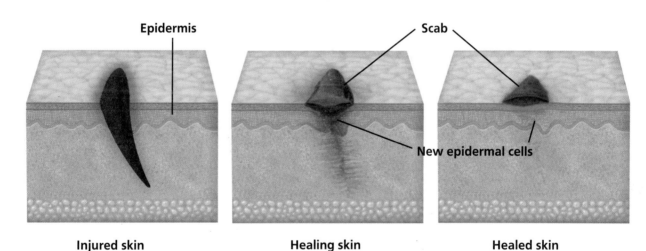

Injured skin   **Healing skin**   **Healed skin**

When there is a cut in the skin's surface, harmful bacteria can enter the body. Your skin quickly replaces the damaged cells with new ones so that the cut is closed up again. It is important to wash even small cuts immediately. This helps your body resist infection.

Sometimes, harmful bacteria survive on the skin. That's why it's important to wash your hands and body often. When you wash, you help get rid of the harmful things.

When you hurt your knee, your skin begins to heal right away. In a few days, new skin cells form under the scab. After a week, the scab is pushed off by the healed skin cells.

Skin helps protect your body from water. It also keeps the inside of your body from drying out. Skin helps you to feel the world around you. Nerve endings in your skin stop you from touching things that are too hot.

The Work That Skin Does

Your skin also helps control your body temperature. When you are hot, sweat glands allow perspiration to escape through the pores in your skin. Along with water, salts are also pushed out through your skin pores. This helps to get rid of wastes from your body.

Finally, the skin is the largest sense organ. Nerve endings in the dermis let you feel heat, cold, pressure, touch, and pain. Along with your other senses (sight, hearing, smell, and taste), the skin gives you information about

When you wash your hands, you help to get rid of harmful bacteria.

your environment and protects you from danger. Your skin is an important organ and needs to be protected so that it can help to protect the rest of your body.

CHECKPOINT
1. Describe the different layers of your skin.
2. How does skin protect your body?
3. What other work does your skin do?
 Why do we need skin?

ACTIVITY

The Sense of Touch

Find Out

Do this activity to learn how the sense of touch can differ on different areas of your skin.

Process Skills

Observing
Communicating
Predicting
Inferring
Controlling Variables

What You Need

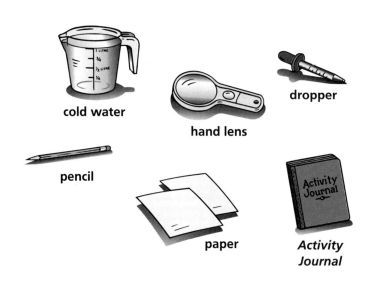

cold water
hand lens
dropper
pencil
paper
Activity Journal

What to Do

1. Use a hand lens to observe closely the skin on the back of your hands, your forearms, and part of your leg. Draw what you see.
2. Predict what will happen when a drop of cold water is put on your hand, your forearm, and your leg. Write your prediction.
3. Have your partner close his or her eyes.
4. Use the dropper to place a drop of cold water on your partner's forearm, on the back of your partner's hand, and on your partner's leg. Be careful to put only one drop on each place.

5. Ask your partner if the water felt the same on each spot. Did the water look exactly the same way on each different part of your partner's skin?
6. **Record** what you observed and what your partner felt.
7. Now trade places with your partner and repeat Steps 3, 4, 5, and 6. **Compare** your findings.

CONCLUSIONS

1. Compare your predictions with your observations.
2. How did you react to each drop of water on your skin?
3. Did the water react the same way each time? Explain.

ASKING NEW QUESTIONS

1. Why did different parts of your skin react differently?
2. **Predict** how the skin on the top of your foot might respond to a drop of cold water. What about the skin on the sole of your foot? Test your prediction.

SCIENTIFIC METHODS SELF CHECK

✔ Did I **observe** my own skin?

✔ Did I **write** my **predictions** and my observations?

✔ Did I try to **explain** why the skin reacted in similar or different ways?

✔ Did I put only **one drop** of water on each spot of my partner's skin?

D11

Skin Controls Body Temperature

Find Out
- How your blood and skin can change your body temperature
- What you can do to protect your skin
- What can damage your skin

Vocabulary
capillaries
sweat glands
goose bumps
sunblock

The Big QUESTION
How does skin control the body's temperature?

There is probably a thermostat in your home or in the school. When it is cold outside, the thermostat triggers a mechanism that turns on the heat. If it gets too warm, the thermostat shuts the heat off. The thermostat regulates the temperature in your home or in the school. Just like the thermostat, your skin helps to regulate the temperature in your body.

Warming and Cooling

The temperature in your body is partly controlled by the flow of blood through your blood vessels near the skin. The blood flowing through your body is warm. Very thin blood vessels in your skin, called **capillaries** (kap′ ə lar ēz), open wider or become narrower in reaction to different brain signals.

When your body gets hot, the brain signals the capillaries to admit more blood so that more blood flows near the surface of the skin. Because the skin on your face is thinner than on other parts of your body, your face can feel and look flushed as you get hot.

This movement of the capillaries helps to cool your body temperature. Blood flows in the capillaries. As more blood enters the capillaries, heat flows from the warm blood through the capillary walls to the skin and then to the cooler environment. This cools the rest of your body. Sweat glands in the dermis also help the body cool off.

Remember that **sweat glands** are tiny tubes that go from inside the body to the surface of the skin. Sweat glands help to maintain your body's temperature. When you get hot, you start to sweat. As sweat evaporates from the skin, your body starts to cool down.

When you get hot, blood and sweat flow near the surface of your skin. This helps to cool your body.

What happens if you jump into a cold pool? You shiver! When your body gets cold, your capillaries become narrow. Warm blood is kept deeper inside the body, away from the surface of the skin, so the body's heat is conserved.

Small hairs on the skin also help control the body's temperature. When it's cold, muscles in the skin make the hairs stand up all over the body. The upright hairs help trap warm air close to the skin. Because the hairs are standing, the surface of the skin looks bumpy. Some people call these bumps **goose bumps.** A layer of fat beneath the dermis also helps keep the body warm.

Protecting Your Skin

Dressing appropriately for the weather can protect your skin and help it work. In cold weather, wearing gloves keeps the skin of your hands from becoming dry and chapped. Wearing a hat helps the head retain body heat when it's cold. More heat is lost from the head than any other body part when it is cold. Wearing a hat reduces this heat loss.

In hot weather, clothes that absorb sweat, like cotton, make you feel more comfortable. If the sweat can't evaporate or becomes trapped on the epidermis, it can cause rashes and body odor. A hat can also shade your skin from the sun on a hot day.

Surfers and scuba divers use special clothing to help control their body temperatures. They wear an extra layer of "skin" when they put on a wet suit. This suit traps warm water next to the skin. The warm water helps to insulate the blood in the capillaries.

A wet suit traps warm water next to the skin. The warm water insulates your blood and helps to keep you warm.

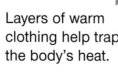

Layers of warm clothing help trap the body's heat.

As you grow older, your skin grows and changes. The sun can damage the layers of your skin over the years, causing it to wrinkle. Smiling, raising your eyebrows, and frowning can all leave small lines or wrinkles on your face. Gravity even pulls down on your skin over the years. Smoking can increase the number of wrinkles on the face and hands. Smoking shrinks the capillaries, preventing nutrients and moisture from reaching the skin. Like the drying from overexposure to the sun, smoking damages the layers of your skin. However, you can also do things to help protect your skin for today and for the future.

Wearing cotton clothing and a hat to shade your skin on a hot day can make you feel more comfortable.

A daily scrubbing with soap and water helps remove dirt, sweat, and germs. A bath or shower removes dead skin cells and helps to keep the skin looking healthy. You can also protect your skin by avoiding overexposure to the sun, wind, and smoke.

If you are going to be in the sun, applying a **sunblock** can shield your skin from the damaging ultraviolet rays. The sunblock filters out some of the dangerous rays. To keep your skin working properly, you need to take good care of it.

A sunblock can help filter out some of the sun's damaging ultraviolet rays.

Ringworm of the scalp is caused by a fungus.

A flea is a parasite that can live on your skin.

Hazards to Your Skin

Lice are parasites that can easily move from one person or animal to another.

Photos are magnified.

Some skin diseases are caused by bacteria, fungi or other parasites. Some fungi cause ringworm and athlete's foot. These fungi can be spread to other people. Someone who is infected with these fungi should not have direct contact with other people's skin because the fungi can be spread. Towels or hats used by infected people can also spread the fungi. Medicines are needed to kill these fungi.

Parasites can also live on your skin. Fleas, ticks, and head lice are small parasites that hop easily from one person or animal to another. Medicines are used to kill fleas and lice because they sometimes cause rashes, sores, or other illnesses. Ticks can be dangerous, and need to be pulled off carefully and then destroyed.

Whenever you ride a bicycle or use in-line skates or a skateboard, you have a chance of falling down and injuring your skin. The best defense against injuring the skin is wearing the proper protective equipment. Wearing knee and elbow pads and a safety helmet will help protect your skin from cuts and scrapes.

Keeping your skin clean, practicing safe habits, and using common sense can all help protect your skin. Three easy steps you can take to protect your skin are: wear a sunblock, avoid the midday sun, and wear proper clothing or equipment. You need to protect your skin so that your skin can continue to protect you.

Wearing safety equipment protects your skin from injury.

CHECKPOINT

1. How does your blood help regulate your body temperature?
2. Name three things you should do to protect your skin.
3. Describe two things that can damage your skin.
4. How does skin control the body's temperature?

ACTIVITY

Facial Wrinkles

Find Out
Do this activity to see how the skin on your face can develop wrinkles as you get older.

Process Skills
Observing
Predicting
Constructing Models
Communicating

What You Need

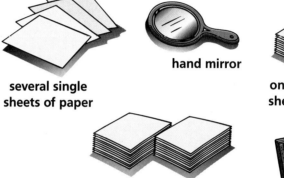

several single sheets of paper

hand mirror

one stack of 10 sheets of paper

one stack of 20 sheets of paper

Activity Journal

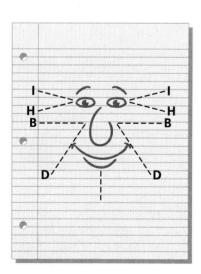

Diagram 1

What to Do

1. Using a hand mirror, observe the skin on your face as you smile and frown.

2. Take a sheet of paper. Draw a face and mark the letters and lines as in Diagram 1. Fold points D and B together. Fold points H and I together. Compare the folds on your paper face with the wrinkles on your smiling face.

3. For a frown, take another sheet of paper. Follow Diagram 2 to draw your face and mark the lines and letters. Fold these points together: A to B, B to C, C to D, I to H, and H to G. Compare the folds on your paper face to the folds on your frowning face.

4. Take a stack of 10 sheets of paper. Try to repeat Step 2 using the whole stack of papers; don't separate them.
5. **Predict** what will happen when you try to fold the stack of 20 sheets of paper. **Write** your prediction. Try to repeat Step 3 with the stack of 20 sheets.

CONCLUSIONS

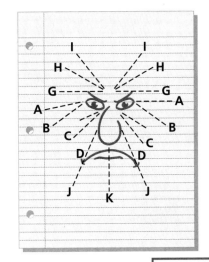

1. Compare your prediction with your observation.

Diagram 2

2. Which paper was easier to fold, the single sheet of paper, the stack of 10, or the stack of 20? Compare what the folds looked like on each of these faces.
3. How can the sheets of paper be compared to the layers of your skin?

ASKING NEW QUESTIONS

1. How do you think the sun or smoking can make the wrinkles in your skin more noticeable?
2. How does your face wrinkle when you are crying? Can you fold your paper to demonstrate this?

SCIENTIFIC METHODS SELF CHECK

✔ Did I **observe** my own face?

✔ Did I **predict** and **write** what would happen as I increased the number of sheets of paper?

✔ Did I try to **model** the folds of skin on my face?

Review

Reviewing Vocabulary and Concepts

Write the letter of the word or phrase that completes each sentence.

1. Your hair, skin, and nails, all of which protect your body and give it its shape, are known as ___.
   - a. integumentary system
   - b. epidermis
   - c. capillaries
   - d. pores

2. We call the material within skin cells that determines how dark or light our skin is ___.
   - a. goose bumps
   - b. sunblocker
   - c. melanin
   - d. sweat glands

3. The thick layer of skin beneath the surface that contains your sweat glands and hair follicles is called the ___.
   - a. epidermis
   - b. sunblocker
   - c. dermis
   - d. capillaries

4. The tiny openings in skin through which sweat is released are called ___.
   - a. capillaries
   - b. sweat glands
   - c. melanin
   - d. pores

Match the definition on the left with the correct term.

5. the thin tubes within the skin through which hairs grow
6. very thin blood vessels in skin that expand or contract in response to different signals such as temperature
7. skin reaction due to hairs standing up in response to cold temperature
8. a lotion which, when rubbed into the skin, can prevent harmful ultraviolet rays from penetrating your skin

   - a. goose bumps
   - b. follicles
   - c. capillaries
   - d. sunblock

D20

Review

Understanding What You Learned

1. What does skin protect the body from?
2. What causes your skin to get flushed when you are hot?
3. Name two things you can do to take good care of your skin.
4. Where do new skin cells begin to grow?

Applying What You Learned

1. What might be the cause of freckles? Explain.
2. Explain why rest rooms in restaurants have signs that read, "Wash your hands with soap before returning to work."

 3. What are the two main things that skin does for your body?

For Your Portfolio

Make a set of flashcards that list possible dangers to the skin, such as sunburn, scraped knee, ringworm, and cold weather. As you look at a card, tell what can cause this danger and what can be done to heal or avoid this danger.

CHAPTER 2
Chemical Substances

Most people try to live healthful lifestyles. We keep our bodies clean by washing and bathing. We brush our teeth, eat healthful foods, get enough rest, and exercise regularly. These are all healthful decisions, and we make them because someone has taught us how to take care of ourselves.

As you grow older, the decisions you need to make may be more complicated than whether to eat apples or candy bars. You may be faced with whether or not to smoke, drink alcohol, or use drugs. What will you do if a friend offers you one of her cold pills when you have a cough? In order to make healthful decisions, you need to learn about chemical substances.

The Big IDEA

Chemical substances affect the body in many ways.

CHAPTER SCIENCE INVESTIGATION

See how to safely use chemical substances found in your home. Find out how in your *Activity Journal.*

LESSON 1

Chemical Substances Cause Changes

Find Out
- What chemical substances are
- What different types of chemical substances there are

Vocabulary
chemical substance
side effect
stimulants
depressants
antibiotic

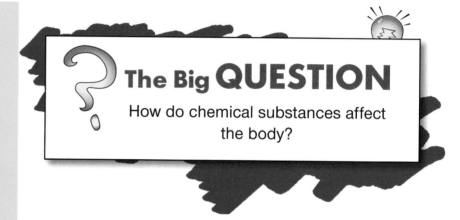

The Big QUESTION
How do chemical substances affect the body?

We often hear about drugs in the news. We hear exciting stories about new drugs that cure diseases and save lives. We also hear about people who misuse, are addicted to, or even die from drugs. So what are you supposed to think about drugs? Are they good or bad? Do you need them, or should you stay away from them?

Chemical Substances

The things that we often call drugs are also referred to as chemical substances. A **chemical substance** is something that changes the way the body operates. Chemical substances may cause positive changes, such as when we take medicine to cure an illness. They may also cause negative changes in our bodies, such as weakening our defense system or damaging our organs.

Sometimes, when a chemical substance is producing a change in one part of the body, it is also causing another change in a different part of the body. The other change, which can be negative or positive, is called a **side effect.**

There are also many different types of chemical substances. It is important to know how each substance affects the body. The more information you have about how chemical substances work, the better you will be able to make decisions about using them. It is also important to talk with your parents and health-care providers about these substances.

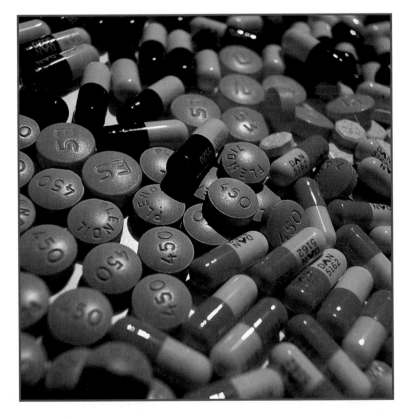

Chemical substances affect your body. They can also cause side effects. You need to know how these substances affect your body so that you can make wise choices.

Types of Chemical Substances

Stimulants

Stimulants (stim′ yōō lənts) are chemical substances that speed up the way the brain works. Caffeine—a drug found in coffee, tea, chocolate, and some soft drinks—is a mild stimulant. Some people buy and use these products to help keep them awake or give them more energy when they are tired.

Stimulants are also often used in medicines or for treating certain illnesses. You can find caffeine in certain headache medicines or cough syrups. A doctor might also prescribe stimulants to treat certain sleep disorders. Some diet pills contain stimulants like caffeine as well.

There are also stronger stimulants that can affect the body in negative ways. These are regulated by the government. Nicotine, found in tobacco products, is one of these.

Before a doctor prescribes a stimulant, or before you use one, it is important to know if the positive uses of the drug outweigh the negative side effects.

Some side effects of using stimulants can be increased blood pressure, headaches, nausea, decreased appetite, irregular heartbeat, or difficulty sleeping.

Depressants slow down the brain. Activities such as driving a car should never be done after taking a depressant.

Depressants

Chemical substances that slow down the brain are called **depressants** (di pres′ ənts). Drinks with alcohol in them are depressants, and so are sleeping pills. Doctors may prescribe depressants to treat people who have difficulty sleeping.

Sometimes when people are using depressants they have a hard time standing or walking. That's because the control of their movements may be affected, and they may be dizzy. Adults who have taken depressants should never drive.

Other side effects include a loss of control over thoughts and speech. People who are taking depressants have to be very careful because it is very easy to become addicted to them. Also, combining alcohol with other forms of depressants can multiply the effect and can cause accidental death.

Antibiotics

Sometimes when you are sick, the doctor may prescribe an antibiotic (ant ī bī ot′ ik). An **antibiotic** is a chemical substance that slows down or kills bacteria that cause diseases. In the past, antibiotics were made from certain molds or bacteria. Today, many antibiotics are created from chemicals.

Antibiotics help the body fight off harmful bacteria. Most bacteria are not harmful, but when harmful bacteria enter the body, they harm the body's cells. Usually the invaders are destroyed by certain blood cells. However, when our defense system is not strong enough to fight off the bacteria, antibiotics are needed to help with the battle.

Penicillin is probably the best-known antibiotic. Penicillin was discovered in 1928 by a Scottish scientist named Alexander Fleming. Patients treated with penicillin recovered quickly and completely. The discovery of penicillin enabled doctors to treat many different diseases and save many lives.

Alexander Fleming noticed that a common fungus killed the bacteria in a dish in his laboratory. He called the fungus *penicillin*.

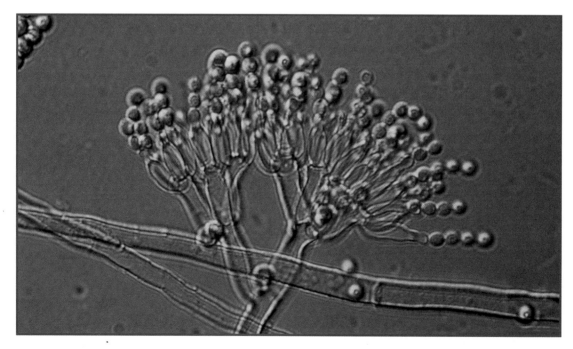

Some people are allergic to certain antibiotics. The allergic reactions can range from mild rashes to stomach problems. In severe reactions, people may even need emergency care. Other people may find that an antibiotic that they used to use no longer helps them fight disease. This usually happens when someone uses an antibiotic often over a long period of time. Luckily, there are many types of antibiotics, so another type can be prescribed by a doctor.

As you can see, different chemical substances can cause different changes in your body. The type of substance and the amount you take can change the effect. Different people may also have different reactions. You are different from everyone else. Your body size and chemistry both influence how a substance reacts in your body. That is why it is so important to talk to your parents or health-care providers about these substances.

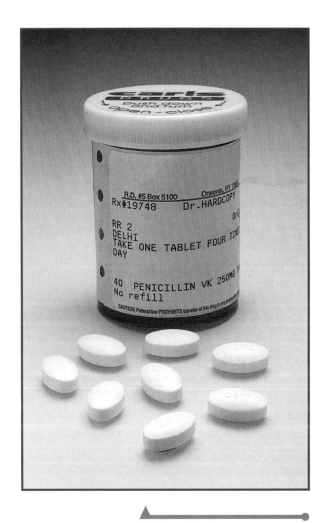

Penicillin is an antibiotic that can be prescribed by doctors.

CHECKPOINT

1. What is a chemical substance?
2. Briefly describe stimulants, depressants, and antibiotics.

 How do chemical substances affect the body?

ACTIVITY
Educating Others

Find Out
Do this activity to teach others about the effects chemical substances have on your body.

Process Skills
Classifying
Communicating

What You Need

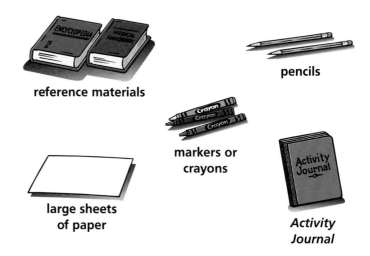

reference materials

pencils

markers or crayons

large sheets of paper

Activity Journal

What to Do

1. Work in small groups. Collect information about stimulants, depressants, and antibiotics. Use your library or the Internet to find other resources if you need to.

2. Write down common examples of each type of chemical substance. For example, coffee and tea contain stimulants, penicillin is an antibiotic, and alcohol is a depressant.

3. List any medical uses for each type of substance.

4. List any possible side effects of each type of substance.
5. With your group, create a separate poster for each substance that illustrates your information.
6. Prepare an oral report about your findings to share with the class.
7. Hang your posters around the school.

CONCLUSIONS

1. Do you think a doctor would prescribe a stimulant for someone who has trouble sleeping? Why or why not?
2. What suggestions could a doctor make to this person before a drug is prescribed? If nothing else works, which type of chemical substance might be prescribed?
3. Explain why someone using alcohol should not drive.

ASKING NEW QUESTIONS

1. Why do some drugs need to be prescribed by doctors only?
2. Why do you think some drugs are illegal?

SCIENTIFIC METHODS SELF CHECK

✔ Did I **classify** chemical substances according to how they affect the body?

✔ Did I **communicate** the information to others?

Positive Effects of Chemical Substances

Find Out
- How prescription drugs are used
- How over-the-counter drugs are used
- How vaccines can prevent disease

Vocabulary
prescription
over-the-counter drug
symptom
vaccine

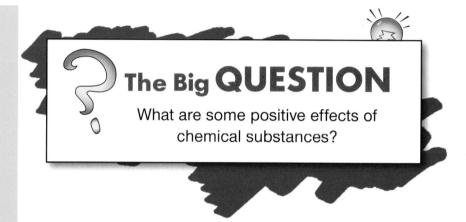

The Big QUESTION

What are some positive effects of chemical substances?

*E*ven though we try to live healthful lifestyles, sooner or later we are bound to end up with some kind of illness. Have you ever missed school because you were sick? If so, you may have been given some medicine to help your body make a positive change. There are four positive effects that chemical substances can have on the body. The effects are curing an illness, helping body organs function, relieving symptoms, and preventing disease.

Prescription Drugs

There are many different infections that can harm the body or make you sick. Strep throat is a common disease caused by germs that infect your body. If you had strep throat, it would be difficult to get better by yourself. You could get plenty of rest and drink lots of fluids.

Someone in your family might even fix you healthful foods, but you might still need some medicine to make you well.

To get well, you could go to see a doctor. After examining you, your doctor may decide that your body's natural ability to heal itself may need help. He or she might write you a prescription for a medicine to cure your illness.

A **prescription** (prē skrip′ shən) is an order written by a doctor for a specific chemical substance. An adult in your family would then take the prescription to a pharmacy, where a pharmacist would get the medicine ready. The pharmacist would get just the right amount for your age and size.

The prescription would have your name on it and give you lots of other information. The prescription would tell you how much of the medicine to take and how often to take it. It might also tell you what kind of side effects the medicine could cause.

After you take the medicine for a few days, you should begin to feel better. If you experience any side effects, you should contact your doctor again. Even if you begin to feel better, you should finish all of your prescription. This is to make sure that all of the germs are killed. By the time the medicine is gone, your strep throat will be gone too. The chemical substance will have cured your illness.

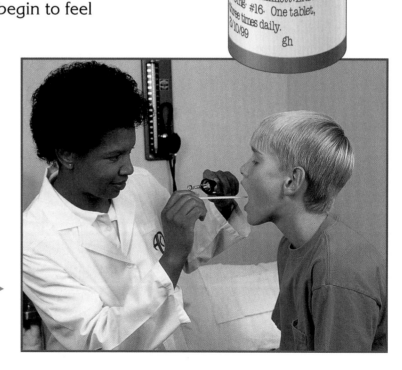

A prescription drug should only be taken by the person whose name is on the label.

A doctor must write a prescription.

Sometimes a prescription drug can help a person with asthma to breathe clearly again.

Your body has many organs that work to keep you alive. Many years ago, if one of your organs was damaged and couldn't do its job, you would have been very sick or died. Today, we have chemical substances that can help damaged organs and keep people alive.

You may know someone who has asthma (az′ mə). More than 10 million people in the U.S. have asthma, a condition that affects the lungs. During an asthma attack, muscles in the lungs tighten and breathing becomes difficult.

For people who have problems with their breathing, there are many prescribed medicines that may be able to help them. Some medicines can reduce the swelling in the lungs. Other prescription sprays can relax the muscles that have tightened around the airways. Chemical substances can help the lungs to function again.

Over-the-Counter Drugs

There are many kinds of medicines that can be bought at the store without a doctor's prescription. These kinds of chemical substances are called **over-the-counter drugs.** Aspirin, cough drops, and cold medicines are types of over-the-counter drugs.

Most over-the-counter drugs will not cure diseases, but they can make sick people more comfortable until they are well. For example, they can help with a cold. Colds cannot be cured. There are no prescription drugs that can cure colds, but there are over-the-counter drugs that can treat the symptoms (simp′ təms) of a cold.

A **symptom** is a sign that you are sick. Symptoms of a cold include a runny nose, a cough, and a sore throat. Over-the-counter drugs, such as cold medicines, cough syrups, or pain relievers, can lessen those symptoms and help you feel better until your body's natural ability to heal itself can take effect. Over-the-counter drugs can relieve symptoms.

It is very important that adults read the label before they buy or take any over-the-counter drug or give it to children. You need to be sure you are taking the right medicine for your symptoms. Just like prescription drugs, you must take the right amount of over-the-counter medicines. If you take too much medicine or take it too often, you may get sick from the medicine. The labels on over-the-counter medicines also warn about any side effects.

Most over-the-counter drugs relieve symptoms. They do not cure illnesses.

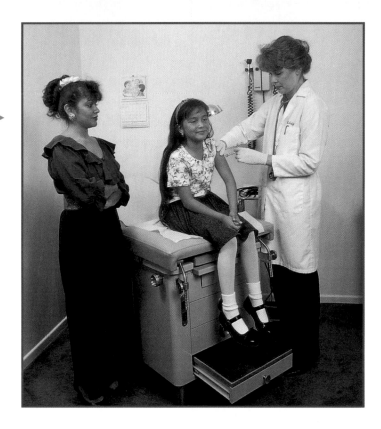

Vaccines can prevent people from getting some diseases.

Vaccines

Wouldn't it be nice if we never got sick at all? Most of us are sure to have a day sick in bed every now and then, but there are many diseases we don't have to worry about getting. Many years ago, people had to struggle with terrible diseases such as polio and smallpox. These diseases were so awful that many people were crippled or died from them.

Today we have vaccines (vak sēnz′) for many different diseases. A **vaccine** is a special chemical substance that protects us against a certain disease. When someone is vaccinated, weak or dead disease germs are injected into his or her body. The injected germs are not strong enough to cause the disease. They cause the body to produce cells to fight off the disease. Recently, scientists have even developed vaccines for chicken pox, some types of flu, and Lyme disease.

Recommended Vaccines

Disease	First doses	Later doses
Diphtheria, Whooping cough (pertussis), tetanus	2 months, 4 months, 6 months, 18 months	DPT vaccine at 4 to 6 years then only tetanus-diphtheria vaccine once every ten years
German measles (rubella)	1 year to early teens	None
Measles	15 months to adult	None, unless first dose was given before 12 months
Mumps	1 year to adult	None
Polio	2 months, 4 months, 18 months	4 to 6 years

You have had many vaccinations since you were born. Scientists and doctors have worked together to come up with a list of suggested vaccinations that all children should have. Some of the diseases you have probably been vaccinated for are measles, mumps, and polio. In most places, you cannot go to school unless you have had all of the recommended vaccinations. Ask a parent to show you your immunization record so you can see what vaccinations you have had.

CHECKPOINT

1. What do prescription drugs do?
2. What do over-the-counter drugs do?
3. How can a vaccine prevent disease?

 What are some positive effects of chemical substances?

ACTIVITY
Reading the Label

Find Out
Do this activity to learn how medicines are recommended for different people depending on their age and size.

Process Skills
Observing
Communicating
Interpreting Data

WHAT YOU NEED

labels from various over-the-counter drugs

Activity Journal

WHAT TO DO

1. Look at and read the labels from different medicines. Answer the following questions.
2. What is the medicine used for?
3. Should adults take this product? If so, how much should they take?
4. Should a five-year-old child take this? If so, how much should he or she take?
5. How often should this medicine be used?
6. What are the possible side effects?
7. Are there any warnings listed on the label? What do they say?
8. What is the medicine's expiration date?

CONCLUSIONS

1. Why should people read the labels on medicines?
2. Why should people know a drug's side effects?
3. Why is it important to check a medicine's expiration date?

ASKING NEW QUESTIONS

1. Are there more over-the-counter medicines for the common cold or for chicken pox? How could you find out?
2. Ask a parent to go through the medicine cabinet at home with you. What do you think you should do with medicines that have expired?

SCIENTIFIC METHODS SELF CHECK

✔ Did I gather information about medicines by **observing** their labels?

✔ Did I **record** my answers?

✔ Was I able to answer the questions by **interpreting** the information on the labels?

Lesson 3

Abuse of Chemical Substances

Find Out
- How alcohol can harm our bodies
- How tobacco harms our bodies
- How illegal drugs can affect our bodies
- About some healthful activities

Vocabulary
drug abuse
addicted
tobacco
marijuana
cocaine
hallucinogens

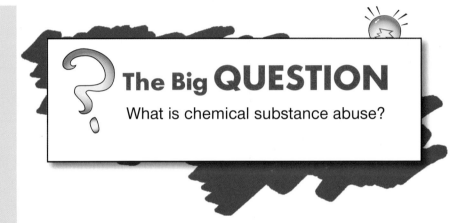

The Big QUESTION
What is chemical substance abuse?

We know that there are drugs that can help make us feel better when we are sick or produce positive changes in our bodies. Other drugs, however, can cause problems for our bodies and for our families.

Alcohol

Sometimes, people use chemical substances when they are not sick. They use them to feel different. When someone overuses drugs that they do not need, it is called **drug abuse,** even if the drug is legal. People who abuse drugs can become **addicted,** which means they can't easily stop using the drugs. Abuse of and addiction to chemical substances can harm your health and hurt your relationships with other people.

Alcohol is a drug found in beer, wine, liquor, and mixed drinks. It is the most widely used drug in the world. It is legal for an adult to use alcohol in the United States. Because alcohol is a depressant, adults sometimes use it to relax. Many people who drink alcohol do not abuse it. They drink responsibly, meaning they do not drink too much. They do not let the alcohol affect their decision making or their behavior. They do not drink alcohol when driving a car.

Alcohol is allowed in some restaurants and at some sporting events, and it is used in some religious ceremonies. Because of this, it is often thought to be less dangerous than other drugs. However, that belief is untrue. Alcohol can be a very dangerous drug.

Unlike other things we eat and drink, alcohol does not have to be digested. It is immediately absorbed into the bloodstream and is carried very quickly to the brain. The alcohol prevents the brain from working correctly, slowing down many of its processes.

People who are under the influence of alcohol may have trouble walking and talking. They may have trouble seeing clearly or remembering things. Heavy drinking can change people's emotions and behavior. Some people become sad. Others may become overly happy or angry. Alcohol can cause people to become loud or even violent. Often they do not make good decisions after drinking.

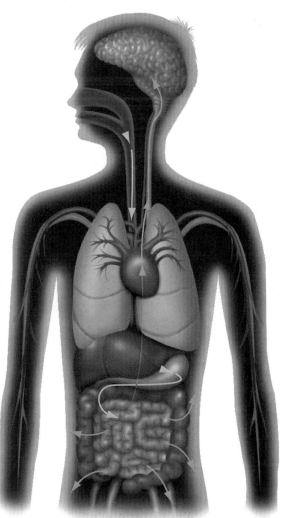

Alcohol is immediately absorbed into the bloodstream and is carried throughout the body. It reaches the brain within minutes.

People who drink regularly risk becoming addicted to alcohol. After a period of time, the brain gets used to the alcohol, so the person needs the drug to feel normal. People who are addicted to alcohol are at risk for many health problems. Heavy alcohol intake increases the risk of liver damage, strokes, high blood pressure, and brain damage. In fact, alcohol actually kills brain cells!

Alcohol can also cause people to seriously damage their relationships with other people. A person who is addicted to alcohol can be more concerned with drinking than with anything else. He or she may have trouble getting along with others or working. Because people who drink too much can get angry and violent, they may harm other people. Many families have been broken apart because of alcohol abuse.

Tobacco

Tobacco (tə ba′ kō) has been said to be the most addictive drug in our society. It is also very deadly. In the United States 8000 people die from smoking-related illnesses every week.

Cigarettes contain more than 4000 harmful substances. Cigarettes are made from tobacco. **Tobacco** is a plant that contains nicotine (ni′ kō tēn′). Nicotine is extremely addictive. Fewer than one out of five smokers who try to quit smoking are able to do so the first time they try. Nicotine is also a deadly poison. If all the nicotine found in a cigar were injected directly into a person, that person would die in less than a minute. When the nicotine is smoked in tobacco, it is less powerful, but over a period of time it is dangerous.

All nicotine products have a warning label.

People who smoke or chew tobacco take in the substances found in tobacco. Nicotine enters the body when the smoke enters the lungs, or it is absorbed in the mouth. It only takes the nicotine less than 10 seconds to travel from the lungs to the brain. There it stimulates the brain and the central nervous system, but afterwards the smoker feels relaxed. People who smoke often do so when they are nervous or upset because smoking relaxes them. But this also makes it very hard to quit smoking.

While the brain is slowing down, the nicotine is making the heart speed up. Nicotine also makes blood vessels smaller, so the heart has to work harder to pump blood through the body. Because a smoker's heart works faster and harder than it is intended to, smokers are at a greater risk for heart attacks.

When tobacco is smoked, many of the poisons join together to form tar. The tar from smoking settles in the lungs and can cause lung cancer. Smoking increases a person's risk for lung cancer.

Scientists have found that smoking can even harm those who do not smoke. People who breathe second-hand smoke can get sick too. Second-hand smoke is the smoke in the air from tobacco products. People who live with smokers can breathe in a lot of second-hand smoke and may end up having health problems too.

Instead of smoking tobacco, some people use chewing tobacco because they think chewing is safer than smoking. However, that is not true. The nicotine in chewing tobacco goes directly into the bloodstream through tissue in the mouth and affects the body in the same way as nicotine from smoking. In addition, people who use chewing tobacco are more likely to get cancer of the mouth and throat.

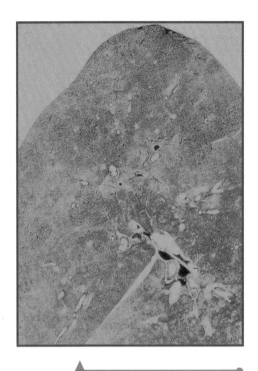

A healthy lung

The tar in cigarette smoke coats the inside of the lungs so they cannot work as they should.

Illegal Drugs

Even though alcohol and tobacco are very dangerous chemical substances, adults can legally buy them. They are sold in almost every neighborhood, at many grocery and convenience stores. Some people also use other drugs that are against the law in the United States. Drugs that people are not allowed to have are called illegal drugs.

Marijuana

Marijuana (mar i wä′ nə), or pot, is a drug that comes from the hemp plant. People have been using it for over 5000 years. Marijuana, like tobacco, can be smoked, and it increases the heart rate. It affects the way a person feels and thinks. Someone using marijuana tends to forget things and finds it difficult to learn new information. When smoking marijuana, some people may even experience a feeling of fear. Because it is smoked, marijuana can also cause some of the same illnesses as tobacco.

Cocaine and Crack

Cocaine (kō kān′) is a very addictive drug from the coca plant. It is a white powder that looks like powdered sugar or baking soda. People sometimes sniff cocaine through their nostrils, smoke it, or inject it into their blood with a needle. Crack is a form of cocaine that is smoked.

The effects of cocaine and crack are extremely unpredictable and dangerous. They can cause seizures in the brain. Using cocaine makes the heart beat even faster than tobacco or marijuana. People have died from heart attacks after using cocaine or crack only one time. Using cocaine or crack can also cause sleeplessness or loss of appetite.

Hallucinogens

Illegal drugs that cause the user's brain to see and hear things that are not really there are called **hallucinogens** (hə lo͞o′ sin ə jəns). These drugs cause the user to hallucinate. Since the user cannot tell what is real and what is not, he or she can act strangely or violently.

PCP and LSD are both dangerous hallucinogens. PCP can block the ability to feel pain, so people can seriously hurt themselves without even knowing it. LSD can make people believe that they have super strength, so they might try things that can result in death. Both LSD and PCP can cause hallucinations long after the drug has been taken.

A normal spiderweb

Making Good Choices

It is true that some chemical substances can cause positive changes in our bodies. However, misuse and abuse of chemical substances are very dangerous. Only you can keep your body free of harmful substances. Looking good and feeling good means saying no to tobacco, alcohol, and illegal drugs.

Finding healthful alternatives can also help you make good decisions. Making healthful choices will help you now and in the future. Healthful choices can include talking to friends and family, getting exercise, playing sports, listening to or playing music, reading books, and writing letters. What other healthful activities can you think of?

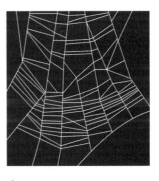

A spiderweb made by a spider affected by a chemical substance

CHECKPOINT

1. How does alcohol affect the brain?
2. How does tobacco affect the heart?
3. How can cocaine affect the body?
4. Name one healthful activity you can do.
 What is chemical substance abuse?

ACTIVITY

Tracing the Effects of Chemical Substances

Find Out
Do this activity to learn how chemical substances can damage your body.

Process Skills
Interpreting Data
Communicating

What You Need

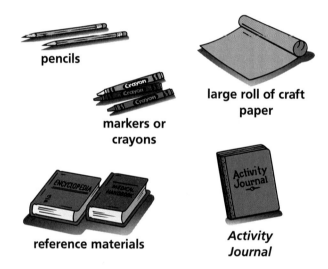

pencils

markers or crayons

large roll of craft paper

reference materials

Activity Journal

What to Do

1. Work in a group. Have your group choose one dangerous chemical substance discussed in the lesson.

2. Use reference books or the Internet to find out how that substance travels through your body. For instance, nicotine first enters your mouth and then travels to your lungs.

3. After researching your substance, have one person lay down on the craft paper. Use a pencil to trace an outline of his or her body. Draw in other body organs, such as the heart, lungs, brain, and so on.

D46

4. Use different-colored markers to trace the path of the chemical substance as it travels through the body. Label and color black every organ the substance comes in contact with.

5. At the bottom of your paper, list all of the effects that the substance can have on the body; for example, increased heart rate, brain seizures, or decreased appetite.

6. Share your information with the class and answer any of their questions.

Conclusions

1. How many different organs are affected by one chemical substance?
2. Which organs of the body are most affected by all chemical substances?
3. Why is it important to know how a chemical substance affects your body?

Asking New Questions

1. Why are some chemical substances illegal?
2. What type of long-term effects can chemical abuse have on a person?

SCIENTIFIC METHODS SELF CHECK

✔ Was I able to answer the questions by **interpreting** the information from our research?

✔ Did I **communicate** by sharing my group's findings with the rest of the class?

Review

Reviewing Vocabulary and Concepts

Write the letter of the word or phrase that completes each sentence.

1. A chemical substance that slows down the way the brain works is called a ___.
   - **a.** stimulant
   - **b.** vaccine
   - **c.** side effect
   - **d.** depressant

2. A chemical substance that slows down or kills bacteria that cause diseases is called ___.
   - **a.** a depressant
   - **b.** a vaccine
   - **c.** a side effect
   - **d.** an antibiotic

3. A chemical substance, such as a pain reliever or cough drops, that can be bought without a doctor's prescription is called ___.
   - **a.** a vaccine
   - **b.** an over-the-counter drug
   - **c.** an antibiotic
   - **d.** a stimulant

4. A chemical substance that protects us against a disease by helping the body make cells to fight that disease is called ___.
   - **a.** a vaccine
   - **b.** an over-the-counter drug
   - **c.** a depressant
   - **d.** a stimulant

Match each definition on the left with the correct term.

5. a chemical substance that speeds up the way the brain works
   - **a.** stimulant

6. a change in one part of the body caused by a chemical substance that is producing a different change in another part of the body
   - **b.** side effect

7. an order for a specific chemical substance written by a doctor and provided by a pharmacy
   - **c.** prescription

Understanding What You Learned

1. Name two common stimulants.
2. What are some side effects of chemical substances?
3. What is it called when someone overuses drugs that they do not need?
4. Name two ways alcohol and tobacco can affect your body.

Applying What You Learned

1. Where can you get the best information about whether you should use a certain chemical substance?
2. Why do we use over-the-counter drugs, rather than prescription drugs, to treat the symptoms of a cold?
3. Describe the differences between an antibiotic and a vaccine.

 4. Explain how chemical substances affect the body.

For Your Portfolio

Make a list of over-the-counter drugs. After each drug, list the symptoms that drug can relieve. With a group of classmates, think about other natural ways you might reduce your symptoms without using the drugs. For example, cough syrup can reduce coughing. Drinking warm tea may also reduce coughing.

CHAPTER 3

Nutrition

Except for the air you breathe, food gives your body everything it needs. Food contains a mixture of nutrients or substances needed to survive. There are six groups of nutrients that you need every day: carbohydrates, fats, proteins, vitamins, minerals, and water. It is important to have a variety of foods in your daily diet so your body can get the nutrients it needs for healthy growth.

The Big IDEA

Nutrients help the body to grow healthy and strong.

CHAPTER SCIENCE INVESTIGATION

Determine how much water is in the foods you eat. Find out how in your *Activity Journal.*

LESSON 1

Carbohydrates, Fats, and Proteins

Find Out
- What carbohydrates do
- Why everyone needs some fat
- Why proteins are important

Vocabulary
nutrients
carbohydrates
fat
cholesterol
proteins

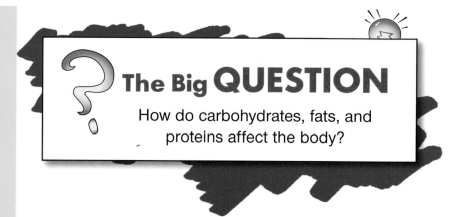

The Big QUESTION
How do carbohydrates, fats, and proteins affect the body?

All day long you are busy working, playing, and relaxing. Even when you are sleeping, your heart is beating and your lungs are breathing air. Where does your body get the energy for you to do these things? You get energy from food.

Carbohydrates

Your body is like a machine. In order to work, it needs fuel. You get fuel from the foods you eat. Your body breaks down food through chewing and digestion so that it can use the parts of food it needs to stay healthy. Substances in food used by the body for energy, growth, and maintenance are called **nutrients** (nōō′ trē ənts). The main source of energy for your body comes from different nutrients.

D52

Carbohydrates (kär′ bō hī′ drāts) are nutrients that provide most of the energy your body needs. Most foods contain carbohydrates in the form of sugars, starches, and fiber. Other than water, carbohydrates are the nutrients you need the most of in your diet.

There are two kinds of carbohydrates: simple and complex. Simple carbohydrates are sugars. They are found naturally in foods like fruit, milk, and some vegetables, such as beets and peas. Sugars from sugarcane and sugar beets are added to food such as candy, soda, cakes, and ice cream as sweeteners. Complex carbohydrates include starches that can be found in foods such as bread, pasta, potatoes, rice, and vegetables. Your body gets its energy from complex carbohydrates by breaking down the food into simple carbohydrates.

Fiber, which is found in some complex carbohydrates, is not a true nutrient, but it is necessary for your body to work. Fiber is the tough part of plants, such as the stringy part of celery. Bran is fiber that comes from wheat and other cereals. Even though your body cannot digest fiber, it helps your digestive system work properly. Scientists believe that fiber helps to protect your body from many diseases.

Sugars are simple carbohydrates.

Your body gets most of its energy from breaking down complex carbohydrates.

Many foods have more than one nutrient. Foods such as potatoes and beans provide carbohydrates in addition to vitamins and minerals.

Prepared foods with a lot of sugar do not have other nutrients. These foods, such as candy and syrup, are said to have empty calories. A baked potato and a can of soda may each have 100 calories. But the soda only gives you sugar. The potato gives you starch, which is broken down in your body to sugar; vitamins; minerals; and even some protein. Which benefits your body more?

When you eat more carbohydrates than your body uses, your body stores them for when you need extra energy. If you don't use them, your body turns the extra carbohydrates into fat. Stored fat is much harder to use as a source of energy.

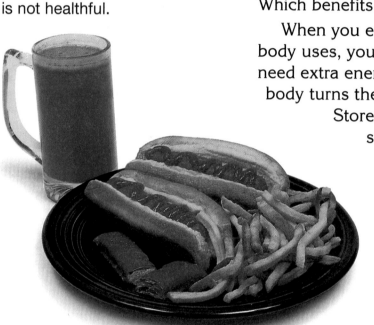

Many people would love to eat this meal. Fat makes food taste good. But too much fat is not healthful.

Fats

Can you imagine eating a whole stick of butter? It doesn't sound very good, but that is how much fat the average person in America eats every day. Americans eat that much fat through foods that contain a lot of fat. You do need a little fat in your diet, but not as much as a stick of butter!

Fat is another way for your body to get energy. All the extra carbohydrates and fats in the body that do not get used up are stored under the skin as a layer of fat.

Fat does do more than just provide you with energy. It carries four important vitamins (A, D, E, and K) throughout your body. Fat helps build nerves and protects many body organs.

You can find fat in both plant and animal sources. Plant fat usually comes in the form of oil. Animal fat is found in butter, cream, egg yolks, and meats. Fat is often added to food during the cooking process. Think about how french fries are cooked. Potatoes are cut and fried in oil. When they are cooking, the potatoes absorb some of that oil.

Cholesterol

Some fat can be digested or broken down into cholesterol. **Cholesterol** (kə les′ tur ol) is a white, waxy substance found in every cell in the body. Cholesterol is a necessary fat. It strengthens cell membranes and protects areas of the brain from harmful substances. Your body makes its own cholesterol in the liver.

Although cholesterol has important jobs to do in the body, too much of it can be dangerous. High levels of cholesterol can build up on the walls of blood vessels, increasing the chances of blood clots, strokes, and heart attacks. People who eat a low-fat diet often have a lower level of cholesterol.

Vegetable oil is an unsaturated fat. What kind of fat do these other foods contain?

Saturated and Unsaturated Fats

Fats can be found in foods in two different forms—saturated and unsaturated. Saturated (sa′ chə rā′ təd) fats are usually solid when they are at room temperature. Unsaturated (un sa′ chə rā′ təd) fats are liquid at room temperature, usually as oil.

Foods high in saturated fats include meat, lard, butter, palm oil, and cocoa butter. People who eat large amounts of saturated fats often have high levels of cholesterol in their blood.

Doctors usually advise people with a high level of cholesterol to use unsaturated fats and not to eat foods with saturated fats. Unsaturated fats, such as olive oil, safflower oil, and canola oil, can decrease the amount of cholesterol in a person's body.

Proteins

It's important to eat foods with protein in them every day. Your body is always building new cells, so you always need protein.

Almost everything in your body is made up of protein. **Proteins** are nutrients that act as the building materials used to repair cells and tissue, and to build new cells. Your body needs many different kinds of proteins. Each protein has its own job to do.

Some proteins are used to build your skin, hair, muscles, teeth, and even your brain. Right now your body is growing. So you need more protein in your diet than an adult does. If you get hurt or injure yourself, your body uses more protein than normal to help repair itself. When you grow older, you will need less protein because your body will stop growing so much.

Proteins help to manage many other processes in your body. They help to fight diseases and to control chemical reactions in your body. Your body can also use proteins for energy; however, the body will use energy from carbohydrates and fats before it will use energy from proteins.

Foods high in complete proteins include eggs, fish, lean meat, and milk. Cereal, grains, beans, peas, nuts, and vegetables can give you some of the proteins you need, but not all. When these foods are combined in certain ways, they can provide your body with the protein it needs. Beans and rice are a good example. Eaten separately, they don't provide enough of the proteins you need. Eaten together, they do.

Your muscles are made of proteins, but you won't build stronger muscles just by eating foods high in proteins. You need to exercise too.

CHECKPOINT

1. What do carbohydrates do?
2. Why does everyone need some fat?
3. Why are proteins important?
 How do carbohydrates, fats, and proteins affect the body?

ACTIVITY

Finding Fats and Starch in Foods

Find Out
Do this activity to see how much fat and starch are in different foods.

Process Skills
Predicting
Observing
Communicating
Defining Operationally

What You Need

tincture of iodine

a variety of foods to test

safety goggles

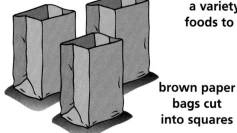

brown paper bags cut into squares

Activity Journal

What to Do

Safety! *Do not taste the food. Throw away all foods tested.*

1. Rub each food on a square of brown paper bag. Let the bag dry.
2. Predict which foods will be high in fat and which ones will be high in starch. Write your prediction.
3. Hold the square up to the light. If fat is present, light will shine through the square. Repeat steps 1 and 3 at least three times to check your findings.

4. Put a drop of the iodine on the food sample.
5. **Observe** the color of the iodine on the food. If starch is in the food, the iodine will turn dark blue.
6. **Record** your observations.

CONCLUSIONS

1. Which foods were high in fat? How can you tell?
2. Which foods were high in starch?
3. Compare your predictions with your observations.

ASKING NEW QUESTIONS

1. How else could you find out if food has fat or starch?
2. Why should you know how much fat is in a food?

SCIENTIFIC METHODS SELF CHECK
- ✔ Did I make a **prediction?**
- ✔ Did I **observe** the changes to the paper and the food sample?
- ✔ Did I **record** my observations accurately?
- ✔ Did I **define** the high-fat foods?

Water, Vitamins, and Minerals

Find Out
- Why water is important for health
- How vitamins and minerals help the body
- What sources are for vitamins and minerals

Vocabulary
vitamins
minerals

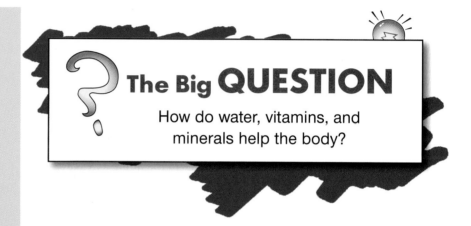

The Big QUESTION

How do water, vitamins, and minerals help the body?

Water, vitamins, and minerals do not give your body energy like carbohydrates, fats, and proteins do. But if you don't have them, you will get sick and die. As long as you eat a variety of foods and drink lots of water, your body will have all the nutrients it needs to stay healthy and strong.

Water and Your Body

You can go for a few weeks without food, but you can only go a few days without water. That is because your body is mostly water. Without water, your body cannot work.

Water in your body has many important jobs. As a part of your blood, it carries nutrients to all your cells and then carries wastes away. Your blood is 90 percent water.

Water helps keep your body working properly. It keeps your body at just the right temperature. Through sweat, your body regulates the heat inside your body and can cool you down when it is hot outside. You lose about 2.5 L of water every day.

Water is easy to replace. Every time you eat food, you take in water. Some foods that contain a lot of water are tomatoes and watermelon. Meat is more than one half water. Bread is one third water.

In addition to getting water from food, it is also important to drink water and other fluids. Milk and juice are nearly all water. Most people need between six and eight glasses of water a day. If you are sick and have a fever, your body will need even more to help regulate your body temperature.

You should drink between six and eight glasses of water a day.

Vitamins

Not all nutrients give the body energy. **Vitamins** help release the energy from carbohydrates, fats, and proteins. Your body only needs a little bit of each type of vitamin. If you don't get enough, you can get sick. If you take too much of certain vitamins, you can also get sick because your body cannot get rid of them.

The best way to get the vitamins you need is through the food you eat. That's because food does more for the body than just give it vitamins. There are about 13 vitamins that are absolutely necessary for good health. Four of them—vitamins A, D, E, and K—are digested with the help of fat in our diets.

The nine other vitamins—the eight B vitamins and vitamin C—aren't stored in your body very long, so you need to eat foods that are good sources of these vitamins every day. The more raw fruits and vegetables you eat, the more vitamins you get. Too much cooking removes most of the vitamins and minerals from foods.

Many vitamins cannot be stored in your body, so you need to eat a variety of foods every day.

Vitamins in Your Food

Vitamin	Needed for	Found in These Foods
A	Good vision, strong bones, healing cuts, healthy skin	Yellow, orange, and green vegetables; yellow fruits; fat of fish, milk, eggs, and liver
B	Using carbohydrates, proteins, and fats in the body; keeping eyes, skin, and mouth healthy; development of the brain	Whole grains, enriched cereals and milk, breads, meats, and beans
C	Healing cuts; development of blood vessels, bones, and teeth; helps minerals to be used by the body	Citrus fruits, melons, berries, leafy green vegetables, broccoli, cabbage, and spinach
D	Works with calcium and phosphorus to build strong bones and teeth; produced in skin getting sunlight	Fatty fish, liver, eggs, butter; added to most milk
E	Helps preserve tissues	Vegetable oils and whole-grain cereals
K	Needed for normal blood clotting; can be made in our bodies	Dark leafy green vegetables, peas, cauliflower, whole grains

Minerals

You only need small amounts of minerals in your diet, but they are essential for your health. **Minerals** are nonliving solids that help to regulate chemical reactions in your body. Minerals can be divided into four groups.

1. ***Minerals that are part of bones.***
   Calcium, magnesium, phosphorus, and fluorine are all important parts of your bones and teeth. You need calcium the most. Calcium makes your bones and teeth strong. It is found in milk products. Magnesium can be found in most foods. Phosphorus is found in peas, beans, broccoli, meat and liver, and whole grains.

2. ***Minerals that control body fluids.***
   Sodium, potassium, and chlorine help keep the right amounts of water inside your body's cells. They help manage the large amount of water in your body. These elements can be found in table salt, spinach, celery, and brown sugar. Potassium can be found in oranges, bananas, and shellfish.

Vitamin C helps your body absorb iron. When you eat whole-wheat toast with orange juice, you help get iron from the toast to the body parts that need it.

3. **Minerals needed to make special materials.** Iron and iodine help your cells do their work. Iron carries oxygen in your blood, among other things. Iodine helps maintain water balance and allows cells to use sugars. Iron can be found in meat, liver, shellfish, peanuts, dried fruits, and eggs. Iodine is in iodized salt and seafood.

4. **Trace elements or minerals needed in tiny amounts.** Trace minerals trigger chemical reactions in the body and are essential for good health. There are 19 trace elements. Zinc and copper are two examples. You can find these two minerals in fish, meat, nuts, grains, and oils.

Milk is an important source of calcium as well as other minerals

CHECKPOINT

1. Why is water an important nutrient?
2. How do vitamins and minerals help the body?
3. What are good sources for vitamins and minerals?

 How do water, vitamins, and minerals help the body?

D65

ACTIVITY
Reading a Food Label

Find Out
Do this activity to learn what nutrients are found in your favorite foods.

Process Skills
Observing
Using Numbers
Interpreting Data

What You Need

food labels from your two favorite foods

Activity Journal

What to Do

1. **Look** at your food labels and **compare** yours to the one shown on the next page. Ask your teacher how to read the **numbers and abbreviations** for information on nutrients.
2. Find all the vitamins and minerals listed on your two labels and **write** these down.
3. List the grams of fat in each food.
4. List the grams of carbohydrates in each food.
5. Compare the numbers from your two labels. How are your foods different?

CONCLUSIONS

1. How are the food labels similar?
2. What is different about the food labels?
3. Why is it important to read food labels?

ASKING NEW QUESTIONS

1. If you have a food allergy, why can food labels be important?
2. Should all people eat the same foods?

SCIENTIFIC METHODS SELF CHECK

✔ Did I **observe** the food labels and find the amount of vitamins and minerals?

✔ Did I **understand** how the numbers on the labels were used?

Review

Reviewing Vocabulary and Concepts

Write the letter of the word or phrase that completes each sentence.

1. Saturated and unsaturated are two different forms of ___.
   - a. fats
   - b. proteins
   - c. cholesterol
   - d. vitamins

2. A white, waxy substance found in every cell in the body is called ___.
   - a. fat
   - b. protein
   - c. cholesterol
   - d. nutrient

3. Nutrients that act as building materials that repair cells and tissue are called ___.
   - a. proteins
   - b. cholesterol
   - c. fats
   - d. vitamins

4. Your body is mostly ___.
   - a. fat
   - b. water
   - c. cholesterol
   - d. protein

5. Nutrients that help release energy from carbohydrates, fats, and proteins are called ___.
   - a. proteins
   - b. carbohydrates
   - c. fats
   - d. vitamins

Match each definition on the left with the correct term.

6. substances in food used by the body for energy, growth, and maintenance
7. nutrients that provide most of the energy your body needs
8. nutrient that carries vitamins A, D, E, and K throughout your body
9. substance that is not a true nutrient but is necessary for your body to work
10. nonliving solids that help to regulate chemical reactions in your body

   - a. carbohydrates
   - b. fiber
   - c. minerals
   - d. fat
   - e. nutrients

Review

Understanding What You Learned

1. What is the main source of energy for your body?
2. Why do doctors recommend limiting the amount of saturated fats we eat?
3. How much water should you drink in one day?
4. What is the function of vitamin D in your body?
5. What two nutrients are needed in the largest amounts?

Applying What You Learned

1. How can eating too many foods high in fat increase your health risks?
2. Why might a doctor advise some people to take extra vitamins?
3. Why should you drink water during and after exercise?
4. Why should vegetables be steamed lightly or eaten raw instead of being overcooked?

5. How do nutrients help the body grow?

For Your Portfolio

Bring in several ads for different brands of the same food product. Look in magazines or newspapers for the ads. What can you tell from the ad about the nutritional content of the food? What can't you tell about the nutritional content from the ad?

Unit Review

Concept Review

1. How does skin protect the body and control the body's temperature?

2. Why is it important to have as much information as possible about chemical substances before you take any?

3. How do carbohydrates, proteins, fats, water, vitamins, and minerals help the body grow?

Problem Solving

1. If somebody is badly burned over a large area of the body, how would the health of the whole body be affected by damage to the skin?

2. Your younger brother has brought a medicine bottle to you. The bottle has your mother's name on it and is labeled *penicillin*. Explain to your brother what the drug is and what it does. Tell him why he shouldn't use it.

3. Scurvy is a disease caused from not having enough vitamin C in your diet. Sailors in the past would often get scurvy because they could not carry foods containing vitamin C on long voyages at sea. Imagine you and your classmates are planning a long trip at sea. How can you make sure that you will have a good source of vitamin C so that you won't suffer from scurvy?

Something to Do

With a group, create a television show called *Is This Your Body?* based on the information in the unit. Pattern your TV show on the game *Jeopardy,* in which contestants get the answers and have to supply the questions. You may even want to create a *Jeopardy* board using posterboard and colored index cards. Then, ask other members of your class or members of other classes that have studied the unit to act as contestants. Award points for correct answers.

Reference

Almanac R2
    Trees R2
    Earth's Geologic Time Table R4
    Landing on the Moon R6
    Rube Goldberg Contraption R8
    Animal Speeds R9
    Rain R10
    Food Guide Pyramid R12
    Your Body R13

Glossary R14

Index R23

Trees

The tallest known tree is a redwood in California that stands 113 meters tall.

The oldest known trees are the bristlecone pine trees, which are thought to be 4700 years old.

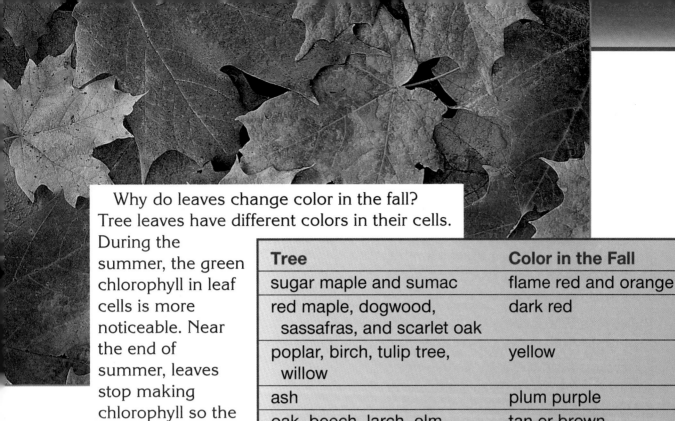

Why do leaves change color in the fall? Tree leaves have different colors in their cells. During the summer, the green chlorophyll in leaf cells is more noticeable. Near the end of summer, leaves stop making chlorophyll so the other colors become visible.

Tree	Color in the Fall
sugar maple and sumac	flame red and orange
red maple, dogwood, sassafras, and scarlet oak	dark red
poplar, birch, tulip tree, willow	yellow
ash	plum purple
oak, beech, larch, elm, hickory, and sycamore	tan or brown

By counting the rings in a tree trunk, the age of the tree can be determined. The size of the rings can help scientists tell what the weather was like years and years ago.

Cloud forests grow on tropical mountain sides where they help to make clouds. The trees are short and gnarled. Nearly every branch is covered with mosses, ferns, lichens, and orchids, which are well adapted to the cold and the mist in the air.

Earth's Geologic Timetable

Precambrian Era is the age of the origin of Earth. Scientists estimate Earth formed 4.6 billion years ago. Types of living organisms during the end of this period included: marine invertebrates, bacteria, algae, and protozoa.

Paleozoic Era is the age of ancient life or the "Age of Fish." Changes in climate and land or water formations resulted in the deposits of salt and coal that we use today. Types of living organisms during this period included: marine animals with shells and skeletons, conifers, insects, club mosses, and ferns.

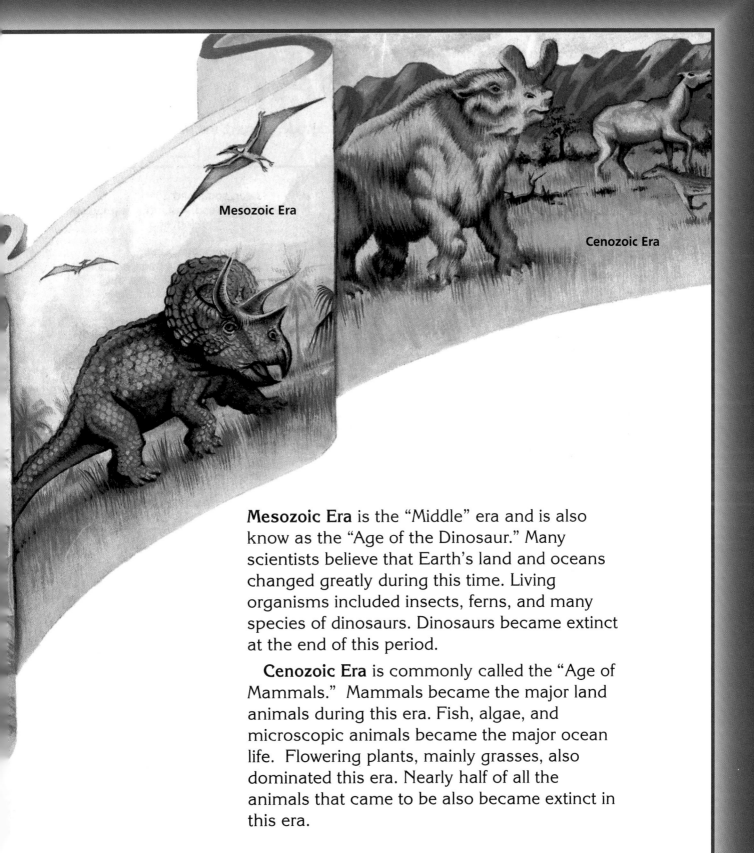

Mesozoic Era

Cenozoic Era

Mesozoic Era is the "Middle" era and is also know as the "Age of the Dinosaur." Many scientists believe that Earth's land and oceans changed greatly during this time. Living organisms included insects, ferns, and many species of dinosaurs. Dinosaurs became extinct at the end of this period.

Cenozoic Era is commonly called the "Age of Mammals." Mammals became the major land animals during this era. Fish, algae, and microscopic animals became the major ocean life. Flowering plants, mainly grasses, also dominated this era. Nearly half of all the animals that came to be also became extinct in this era.

Landing on the Moon

Because of the moon's quiet and still environment, there are human footprints in the dust on the moon that won't fade for centuries. Transporting a human there to make those footprints is one of the greatest human achievements. Making the dream of putting a human on the moon a reality took many years.

The first humans to visit the moon were Neil A. Armstrong and Edwin E. Aldrin, Jr. They landed the lunar module named *Eagle* in the moon's Sea of Tranquility. The first words sent from the moon were from Armstrong who radioed: "Houston, Tranquility Base here. The *Eagle* has landed."

On July 20, 1969, Neil A. Armstrong made "one giant leap for mankind" as he stepped on the moon.

Because of the difference in gravity, a golf ball on the moon would be about six times larger than a golf ball of the same weight on Earth.

Aristotle concluded from the shape of Earth's shadow during a lunar eclipse that Earth must have a round shape.
340 B.C.

Chinese developed gunpowder rockets.
1232

Nicolaus Copernicus insisted that the planets all travel around the sun. Few people believed him.
1543

Humans have brought back 300 kg of moon rocks. Scientists have found those rocks are much like Earth rocks. Some are dark rocks called basalts, which are like rocks found in Hawaii. Scientists have concluded that Earth and the moon formed at the same time, about 4.5 billion years ago.

Space Facts

- Humans have visited the moon seven times.
- Astronauts in space grow 3.5 to 5.5 centimeters taller. This is because without Earth's gravity, their spines lengthen and straighten out.
- Gravity on the moon is about 1/6 of the gravity on Earth.
- Other planets also have moons. Mars has 2 moons. Jupiter has at least 16; Saturn has 18; Uranus has 15; Neptune has 8; and Pluto has 1.

American astronomers took the first photos of the moon.
1840

Soviet Yuri A. Gagarin became the first human in space.
1961

Apollo 11 astronauts from the United States became the first to land on the moon.
1969

The first US space shuttle, *Columbia*, took off.
1981

The Soviets launched the first part of the space station *Mir*.
1986

The first part of The International Space Station was launched.
1998 2000

Rube Goldberg Contraption

Reuben Goldberg (1883-1970) was a draftsman and a cartoonist who made up bizarre inventions that mixed silliness with seriousness. Can you follow the workings of this contraption?

Animal Speeds

Most animals would not have any problems staying under the speed limit. Other animals rely on their great speed to capture prey or to escape from predators.

Animal	Speeds*
dragonfly	50 kph
racing pigeon	85 kph
peregrine falcon	180 kph when diving
brown hare	26 kph
ostrich	72 kph
antelope	89 kph
cheetah	105 kph
gentoo penguin	27 kph
sei whale	48 kph
tunny fish	71 kph
giant tortoise	.25 kph
daddy-long-legs	1.62 kph

*listed in kilometers per hour

Ways Animals Move

Hummingbirds can use their wings to hover in midair. Hovering uses up more energy than flying because the wing muscles have to beat very quickly.

Snails creep along on a single foot that acts like a sucker. They also let out a slimy mucous trail that helps them to move along rough surfaces.

Rain

What is the shape of a raindrop?

Raindrops are not really pear-shaped or tear-shaped. High-speed photography shows that air pressure flattens the bottom out so the raindrop is donut-shaped with a hole almost through the center.

Where does the rain fall?

The place with the most rainy days per year is Mount Waialeale on Kauai, Hawaii. It has up to 350 rainy days per year.

What does lightning look like?

Lightning is usually about 1.6 kilometers long but it can be up to 32 kilometers long. Lightning can travel from 160 to 1600 kilometers per second.

Lightning can heat up the air temperature along its path to 30,000° C. That is six times hotter than the surface of the sun.

Lightning can appear as zigzags, forks, sheets, ribbons, chains, and even balls. Ball lightning is rare but it has been described as a moving, colored circle that can last for a few minutes. St. Elmo's fire has been described as a form of electric discharge that occurs around high, grounded metal objects like chimneys and ship masts. It gets its name from sailors who first saw the flame-like colors on their ship masts. St. Elmo is the patron saint of sailors so they named the electric discharge after him.

How far away is lightning?

When you see a lightning flash, count the number of seconds until you hear the sound of thunder. Divide the number by five to find out how many miles away the lightning was.

R11

Food Guide Pyramid

A Guide to Daily Food Choices

Fats, Oils, and Sweets
Use Sparingly

Milk, Yogurt, and Cheese Group
2–3 servings

Meat, Poultry, Fish, Dry Beans, Eggs, and Nuts Group
2–3 servings

Fruit Group
2–4 servings

Vegetable Group
3–5 servings

Bread, Cereal, Rice, and Pasta Group
6–11 servings

Your Body

Fingerprints

The pattern on the tips of your fingers makes your fingerprints. Since no two people have the same pattern, fingerprints can be used to identify you. Your skin can replace itself each month, but your fingerprints remain the same for your whole life.

Hair Types

Hair grows on every part of your skin except the soles of your feet and the palms of your hands. Just like skin, your hair color is determined by melanin. The shape of your hair can also determine how it looks. If you cut a strand of hair crosswise, curly hair would look square. Wavy hair would look like an oval, and straight hair would be round.

Fingernails and Toenails

Your nails grow from the fold of skin at the nail's root. Like your hair, your nails are dead, but the skin at the base of the nail can sense touch and pressure.

Your fingernails grow about 2 cm a year. Fingernails grow four times as fast as toenails.

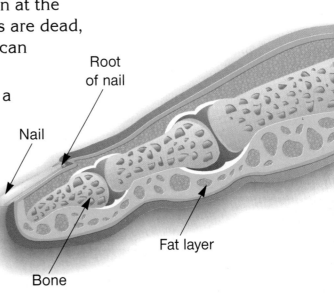

R13

Glossary

A

abyss the deepest areas of the ocean

acceleration the rate at which the speed of an object changes

adaptation a feature, behavior, or characteristic that helps a species survive in its environment

addicted when someone cannot stop taking drugs

aftershocks smaller earthquakes following and radiating from the center of an earthquake

air mixture of invisible gases, mainly oxygen and nitrogen, that surrounds Earth

air mass a large body of air with nearly the same temperature and moisture throughout

air pressure the weight of air pushing down on Earth's surface

antibiotic a chemical substance made from certain molds or bacteria and designed to kill germs that cause diseases

artificial satellites objects built by humans and launched into space with rockets to orbit a larger body in space

avalanche a sudden downward movement of snow and ice

B

barrier anything that stops migration from one area to another

buoyancy the ability of an object to float or rise in a liquid

C

capillaries thin tubes in the body that carry blood between the arteries and veins

carbohydrate a nutrient that provides most of the energy your body needs

carbon dioxide a colorless, odorless gas made up of carbon and oxygen

carbon dioxide–oxygen cycle the ongoing exchange of carbon dioxide and oxygen between plants and animals

cartilage a tough, flexible material that helps to support and shape body parts

chemical substance a drug that changes the way the body works

chemical weathering the process in which minerals that make up rocks undergo chemical changes

Glossary

chlorophyll a green pigment in plants that helps them make food by absorbing energy from light

cholesterol a white, waxy substance found in every cell in the body; it helps break down fats and strengthens cell membranes

circuit a path for electrical energy to move through

circuit breaker a device that stops the flow of current before wires in a circuit become too hot

clay a kind of soil that can be molded into pottery and bricks when moist and that turns hard when dried

cocaine a fine, white powder that is a dangerous drug derived from the coca plant

cold front when a cold air mass slides under a warm air mass, producing fast-rising air and strong winds

cold-blooded the ability of living things to change their body temperatures to their surroundings

combustion a chemical reaction caused by burning

competition the struggle of one organism against another to get what it needs to live

composting mixing organic matter, such as grass clippings, shredded branches, and food scraps, with water and letting it decay to make a rich fertilizer

condensation the change of a gas to a liquid

conduction the transfer of heat from atom to atom and molecule to molecule

consumer an organism that can't make its own food and therefore eats plants or other consumers

convection the transfer of heat by the movement of heated liquids and gases

crystals solids composed of regular arrangements of particles; the building blocks of minerals

D

dark zone the zone in the ocean where there is no light, the water is severely cold, and the water pressure is severe

decomposer a consumer, such as mold, that gets its energy by causing dead organisms to decay

density a measure of how much mass an object has for its volume

deposition the buildup of particles moved by erosion

depressant a chemical substance that slows down the brain

Glossary

dermis the thick layer of skin found beneath the epidermis

drug abuse the overuse and improper use of drugs

E

earthquake a sudden movement of Earth's crust caused by a release of energy built up along a fault

ecosystem all the living and nonliving things within an area and their relationships with each other and their physical environment

electric conductor a material, such as gold, silver, copper, or aluminum, that electric current flows through easily

electric current the flow of electrons along a path

electric field the space around a charged object that pushes or pulls on another charged object

electromagnet an arrangement of iron wrapped in a coil of wire carrying an electric current

electron a negatively charged particle

energy pyramid a model that shows the flow of energy from the bottom to the top of a food chain

epidermis the thin outer layer of skin

erosion the process by which weathered particles, such as pebbles, sand, and dust, are carried away and deposited somewhere else by wind, water, gravity, or ice

evaporation the change of a liquid to a gas

F

fat an oily substance found in animals, some plants, and certain foods that gives us energy and is stored in the body to provide warmth

food web a structure of overlapping food chains

force a push or pull that one object has on another object

fossil fuel a carbon-rich fuel, such as coal, oil, or natural gas, formed from the remains of ancient animals and plants

fossils traces or remains of plants or animals that are formed when the hard parts of plants and animals are preserved in mud that turns to rock

freezing the changing of a liquid to a solid

Glossary

friction a force that opposes motion between two surfaces that are touching each other

front the place where two different air masses meet

fuels sources of energy that are burned for heat or energy

fuse a safety device that stops the flow of current before wires become too hot

G

geologic timetable a record of four eras in Earth's history: Precambrian, Paleozoic, Mesozoic, and Cenozoic; large extinctions of plant and animal life occurred at the end of each era

germination the process of growing shoots and roots that seeds and beans go through

glacier a large mass of slowly moving ice

goose bumps the bumpy surface of the skin when small hairs stand up and trap warm air close to the skin, helping to control the body's temperature

gravitational pull the force objects exert on other objects

grounding transferring extra electrical charge to the ground

H

hair follicles thin tubes through which hairs grow

hallucinogens drugs that cause the user to see and hear things that are not really there

horizons what scientists call layers of soil

humus decayed plant or animal matter

hurricane a violent storm that forms from warm, moist air over an ocean

I

inclined plane a flat surface that is higher at one end

index fossils the remains of animals and plants that lived for a brief amount of time and that can be used to help scientists determine the age of rock layers

inertia the resistance of a body to a change in its motion

inexhaustible resource the things we use that can never run out, such as sunlight

inherited traits characteristics that are passed from an adult to an offspring

instincts behaviors that animals are born with

Glossary

insulators materials, such as rubber, air, and plastic, that electric current or heat does not easily move through

integumentary system a system in the human body that is made up of skin, hair, and nails

invertebrate an animal without a backbone

L

landslide a quick movement of soil or rock down a slope

larva an insect at the stage of development after it is an egg and before it becomes a pupa

lava molten rock (magma) that is released at Earth's surface during volcanic eruptions

learned behavior behavior that is changed by experience

lever bar that can turn on a fixed point

life cycle the series of changes each living thing goes through from life to death

luster the physical property that describes the way a mineral reflects light, such as bright and shiny or dull and pearly

M

machine a device that performs work

magnetic field the space around a magnet that pushes or pulls on objects containing iron or other magnets

magnetic poles the places on a magnet where its attraction is the strongest

magnetized what happens to a piece of material that has been made into a magnet

marijuana a drug derived from the hemp plant and smoked like tobacco

mass the amount of matter in an object

matter anything that takes up space and has mass

melanin the dark brown pigment that determines the color of hair and skin

metamorphosis the change in form or function of a living organism by natural growth or development

migration moving to a new area that has more room and resources

Glossary

mineral a natural, nonliving solid made of one or more of Earth's elements, such as silicon, oxygen, carbon, and iron; in health, minerals help to regulate chemical reactions in your body

N

niche the status or role of an organism within its community

nonvascular plant a plant that absorbs water through its surfaces

nutrients substances used by organisms for energy, growth, and maintenance

O

orbit the path an object follows when it revolves around another object

over-the-counter drug medicine that can be purchased at the store without a doctor's prescription

oxygen an odorless, colorless gas that is part of the air we breathe

P

parallel circuit a circuit in which the current flows through different parts of the circuit at the same time

photosynthesis the process by which plants use energy from light to produce food

physical weathering the process by which wind, water, or ice reduces rocks to smaller pieces without changing their composition

phytoplankton tiny plantlike organisms that float at or near the water's surface

pioneer stage the first stage of change in succession, in which grasses, other plants, rodents, birds, and insects grow and live

planets large objects in space that revolve around the sun

pores openings in the skin's epidermis

precipitation the falling of water from the sky in the form of rain, sleet, hail, or snow

prescription an order written by a doctor for a specific chemical substance

producer a plant that uses the sun's energy, along with other substances, to make its own food

protein a nutrient that helps repair cells and tissue and build new cells

Glossary

protons positively charged particles in the nucleus of an atom

pulley a wheel or wheels around which a rope fits to lift or lower heavy loads

pupa an insect at the stage of development after it is a larva and before it becomes an adult

R

radiation the transfer of heat in waves through space

rain shadow the land that gets little rain on the side of the mountain away from the ocean

relative dating figuring out the order in which events occur by looking at the layers of rock

resistors a material, such as the wire in a lightbulb, that doesn't conduct electric current very well

rock cycle the set of processes that changes rocks, such as forming, wearing down, and re-forming

rockets tube-shaped vehicles with expanding gases that can escape from only one end

S

sand a kind of soil made of tiny grains of rock

satellite an object that revolves around a larger body

screw an inclined plane that winds around a rod into a spiral

sediment loose material from erosion and weathering

series circuit an electric circuit in which the parts are connected so that the current flows through each part of the circuit, one after another

short circuit an excessive current flow that often causes damage

side effect an additional change in the body caused by taking a chemical substance designed to produce another kind of change

silt a kind of powdery, slippery soil

soil a mixture of weathered rock, decayed plant and animal material, water, and air

soil conservation the preservation and protection of land

solar system the name for our sun and all the objects in orbit around it

Glossary

space probes rocket-launched vehicles that carry data-gathering equipment to distant locations in space

space shuttles reusable crafts designed to transport astronauts, materials, and satellites to and from space

species a group of organisms that usually reproduce only with other organisms of the same species

speed the measure of how quickly an object moves over a certain distance

spore a plant cell that develops into a new plant

static electricity a form of electricity in which electric charge does not move

steam an invisible, gaseous form of water

stimulant a chemical substance that speeds up the way the brain works

streak the powder left when a mineral is rubbed against a hard, rough surface

succession the process by which new plant and animal populations replace old plant and animal populations over time

sunblock a lotion that keeps harmful ultraviolet rays from penetrating the skin's epidermis and causing sunburn

sunlit zone the upper layer in the ocean where sunlight shines and seasonal changes influence the water temperature

sweat glands tiny tubes that go from inside the body to the surface of the skin to cool off the body

switch a device used to open and close circuits without disconnecting wires or unscrewing lightbulbs

symmetry the way that an animal's body parts are arranged

symptom a sign that you are ill

T

telescopes instruments for viewing distant objects, such as planets, moons, and stars

thawing the changing of a frozen solid to a liquid

thunderhead a tall, dark cloud that can produce lightning, thunder, and heavy rains

tides the rise and fall of the ocean's surface

tobacco a plant that contains nicotine, an extremely addictive drug found in cigarettes and cigars

Glossary

tornado a dark, whirling, funnel-shaped column of air that moves rapidly and often destroys things in its path

transpiration the process of plants releasing water from their leaves during photosynthesis

twilight zone the zone in the ocean below the sunlit zone where the water is colder and the water pressure is greater

V

vaccine a special chemical substance that protects us against a specific disease

vascular plant plant that moves water through tiny tubes from the roots to the stems and leaves

vertebrate an animal with a backbone

vitamin a substance in food that helps our bodies stay healthy

volcano an opening in Earth's crust through which steam, other gases, lava, cinders, and ashes erupt

volume the amount of space matter takes up

W

warm-blooded the ability of vertebrates to maintain the same body temperature regardless of their environment

warm front when a warm air mass slowly rises over a cold air mass, resulting in gentle rain

water cycle the repeating cycle of evaporation, condensation, and precipitation that occurs on Earth

water retention the ability of soil to hold water

water vapor the gas produced when water evaporates

wedge an inclined plane used to push things apart

weight the amount of force that gravity exerts on an object

wheel and axle a simple machine with a wheel that turns a post called an axle

wind the movement of air from higher-pressure areas to lower-pressure areas

Z

zooplankton tiny, animal-like organisms that float at or near the water's surface

Index

A

abyss (ocean), A72
acceleration, C94–C95
activities
   Adding to Ecosystems, A82–A83
   Animal Behavior, A26–A27
   Classifying Rocks, B98–B99
   Comparing Animal Life Cycles, A38–A39
   Comparing Density, C88–C89
   Decomposers, A74–A75
   Educating Others, D30–D31
   Eroding with Water, B78–B79
   Facial Wrinkles, D18–D19
   Finding Fats and Starch in Foods, D58–D59
   Food Webs, A54–A55
   Gravitational Pull, B20–B21
   Ice Power, B70–B71
   Influencing an Ecosystem, A66–A67
   Investigating Vertebrates, A18–A19
   Magnetic Fields, C18–C19
   Make a Bulb Light Up, C32–C33
   Making a Balloon Rocket, B28–B29
   Making a Barometer, B58–B59
   Making a Cloud, B50–B51
   Making an Earthquake, B86–B87
   Making Circuits, C40–C41
   Making Magic with Air, B40–B41
   Modeling Earth's Limited Resources, B122–B123
   Modeling the Solar System, B12–B13
   Moving Water Through Plants, A10–A11
   Naming Minerals, B106–B107
   Observing Heat Transfer by Conduction, C68–C69
   Observing Inertia, C96–C97
   Observing Transpiration, A46–A47
   Opposites Attract, Likes Repel, C10–C11
   Producing Heat, C60–C61
   Reading a Food Label, D66–D67
   Reading the Label, D38–D39
   Sense of Touch, The, D10–D11
   Testing Soil Characteristics, B114–B115
   Tracing the Effects of Chemical Substances, D46–D47
   Turning a Magnetic Field On and Off, C48–C49
   Using Heat to Do Work, C76–C77
   Using Simple Machines, C106–C107
adaptation
   behavior, A24–A25
   migration, A64
   mimicry, A23
   ocean life, A68–A69, A73
   structural, A22–A23
addicted, D40
aftershocks, B82
air, B34–B39
air mass, B52–B53
air pressure, B35
alcohol, D40–D42
animals
   behavior, A24–A25
   carnivores, A16

Index

classification of, A12–A17
cold-blooded, A15
herbivores, A16
in oceans, A68–A73
instincts, A24
invertebrates, A13
life cycles, A34–35, A36–A37
metamorphosis, A34–A35
mimicry, A23
omnivores, A16
symmetry, A17
vertebrates, A14–A15
warm-blooded, A14
antibiotics, D28
aquatic ecosystems, A58, A68–A73
artificial satellites, B24–B25
atom(s), C5
avalanches, B80–B81
axle and wheel, C102

B

barriers
migration, A65
behavior
instincts, A24
learned, A25
buoyancy, C87

C

capillaries, D12–D13
carbohydrates, D53–D54, D62
carbon dioxide, A4, A40–A41, A44–A45, C57
carbon dioxide-oxygen cycle, A41, A44–A45
carnivores, A16
cartilage, A14
cells
animal, A12
plant, A4, A5, A7, A32
skin, D4–D5, D7
chemical substance(s)
abuse, D40–D45
alcohol, D40–D42
antibiotics, D28–D29
cocaine, D44
crack, D44
definition, D24
depressants, D27
hallucinogens, D45
healthful choices, D45
illegal drugs, D44–D45
marijuana, D44
over-the-counter drugs, D35
prescription drugs, D32–D34
side effects, D25–D26, D35
stimulants, D26
symptom, D35
tobacco, D42–D43
vaccines, D36–D37
chemical weathering, B65, B67
chlorophyll, A5
cholesterol, D55–D56
circuit, electrical
circuit breakers, C38–C39
closed, C26–C27
diagrams, C30–C31
fuses, C38–C39
open, C26

Index

parallel, C36–C37
series, C34–C35, C37
short circuit, C39
cirrus clouds, B47
classification
of animals, A12–A17
of plants, A8–A9
clay, B111
climate, B69
cloud(s)
formation, B43–B44
thunderheads, B54
types, B46–B47
cocaine, D44
cold-blooded, A15
cold front, B55
combustion, C57–C59, C74
compasses, C13
competition, A62–A63
composting, B121
condensation, B43–B44
conduction, C64–C65, C67
conductors, C28
electric, C28
heat, C64–C65
conservation of soil, B120–B121
consumers, A50, A52
convection, C65–C66
crystals, B101
cumulus clouds, B46
cycles
carbon dioxide-oxygen, A41, A44–A45
life, A32–A37
rock, B94

water, A42–A43, B42–B49

D

dark zone (ocean), A72–A73
decomposers, A51, A70, A77
density, C86–C87
deposition, B73
and erosion, B73
depressants, D27
dermis, D6
desert, B48
drug abuse, D40–D45
drugs. *See* chemical substances

E

Earth
and the moon, B14–B19
gravitational pull, B5, B14
orbit, B5
surface changes, B62–B69, B72–B77, B80–B85
earthquake(s)
aftershocks, B82
definition, B82
measuring, B83
predicting, B84
ecosystem(s)
changing, A76–A81
characteristics of, A60–A65
competition in, A62–A63
humans in, A44–A45, A65, A80
niche, A63
ocean, A68–A73
succession, A78–A79

R25

Index

electrical energy
   conductors, C28
   electric current, C24–C31
   electric field, C7
   electrical charge, C5–C9
   electrical circuits, C25–C27, C30–C31, C34–C39
   electrical switch, C27
   electromagnets, C42–C47
   electrons, C5–C9
   generating, C72–C73
   grounding, C9
   insulators, C28
   protons, C5, C7
   resistors, C29
   static electricity, C4–C5, C8–C9, C24
electromagnet, C42–C47
electrons, C5–C9
elements, B100
energy
   flow, A48–A53
   fuels, C54, C57
      in food webs, A49–A53
      in human body, D52–D54, D56
   pyramid, A52–A53
   wind, B38–B39
epidermis, D5
erosion
   agents of, B73–B77
   definition, B72
   deposition, B73
   sediment, B74–B76
evaporation, B43, B44
extinction, A80

F

fats, D54–D55, D62
ferns, A9
fingerprints, R13
Fleming, Alexander, D28
flowers, A6, A9
Food Guide Pyramid, R12
food webs, A49
force
   and motion, C90–C95
   and work, C99–C105
   friction, C92–C93
   of erosion, B72–B73
fossils, B96–B97
freezing, B66
friction, C92–C93
fronts, B53–B55
fuels, C54, C57
   fossil, A44, C58–C59
   gasoline, C74
fuses, C38–C39

G

geologic timetable, B97, R4–R5
germination, A33
Gilbert, William, C5
glacier(s), B73–B74
goose bumps, D13
gravitational pull
   and air, B35
   definition, B5
   of moon and Earth, B14–B19
grounding, C9

Index

H
hair follicles, D6
hair types, R13
hallucinogens, D45
heat
    combustion, C57–C59, C74
    conduction, C64–C65, C67
    convection, C65–C67
    definition, C55
    fuels, C54, C57
        fossil, C58–C59
    measuring, C56
    production, C54–C59
    properties, C55, C62
    radiation, C63
    transfer, C62–C67
herbivores, A16
horizons, soil, B117
humus, B109–B110
hurricane, B56

I
igneous rocks, B92–B94
inclined plane, C104
index fossils, B96
inertia, C90–C91
inexhaustible resources, B120
inherited traits, A21, A24
inner planets, B6–B7
instincts, A24
insulators, C28, C67
integumentary system, D4, R13
invertebrates, A13

L
landslides, B80–B81
larva, A34
lava, B85, B92–B93
learned behavior, A25
leaves, A5, A8, A22, A42, R3
lever(s), C99–C101
life cycles
    animals, A34–A35, A36–A37
    plants, A32–A33, A36
lightning, B54, C8–C9, R11
luster, B103

M
machines
    simple, C98–C105
magnet(s), C13–C17, C42–C47
magnetic field, C15
magnetic poles, C14–C15
magnetism
    discovery of, C12–C13
magnetized, C16
marijuana, D44
mass
    and matter, B35, C84–C85
    of air, B52–B53
matter
    buoyancy, C87
    definition, C80, C82–C83
    density, C86–C87
    mass, C84–C85
    properties of, C82–C87
    volume, C83
    weight, C85

Index

melanin, D5–D6
metamorphic rocks, B93
metamorphosis, A34–A35
migration, A64–A65
mimicry, A23
minerals
    as nutrients. *See* nutrition
    characteristics of, B100
    crystals, B101
    definition, B100
    identifying, B102–B103
    luster, B103
    streak, B103
    uses of, B104–B105
Mohs, Friedrich, B102
moon
    and Earth, B14–B19
    and tides, B18–B19
    gravitational pull, B17
    landing on the, R6–R7
    orbit, B16
    See also solar system
moonlight, B15
mosses, A9
motion
    acceleration, C94–C95
    and force, C90–C95
    inertia, C90–C91
    speed, C93

N

natural resource(s)
    fossil fuels, C58–C59
    inexhaustible, B120
    soil, B116–B121

niche, A63
nonvascular plants, A9
nutrients
    definition, D50, D52
    in ecosystems, A4, A5, A70, A77
nutrition
    carbohydrates, D52–D54, D62
    cholesterol, D55–D56
    fats, D54–D56, D62
    minerals, D64–D65
    proteins, D56–D57, D62
    vitamins, D62–D63
    water, D60–D61

O

ocean(s)
    currents, A69, A71
    decomposers, A70
    ecosystems, A68–A73
    layers (zones), A72–A73
    minerals in, A71
    phytoplankton, A70
    tides, B18–B19
    waves, B77
    zooplankton, A70
omnivores, A16
orbit, B5, B10–B11, B14–B17, B25
organisms, A2, A20
outer planets, B8–B9
over-the-counter drugs, D35
oxygen, A5, A40–A41, B35

Index

P

parallel circuit, C36–C37
photosynthesis, A5, A41, A49, A70
physical weathering, B65
phytoplankton, A70
pioneer stage (succession), A78
plane, inclined, C104
planet(s)
    days, B10–B11
    Earth, B7
    Jupiter, B8
    Mars, B7
    Mercury, B6
    Neptune, B9
    Pluto, B9
    Saturn, B8
    Uranus, B9
    Venus, B6
    years, B10–B11
plant(s)
    and erosion, B119–B121
    carbon dioxide, A4–A5, A41, A44–A45
    cells, A4–A7, A32
    chlorophyll, A5
    classification of, A8–A9
    cones, A7
    ferns, A7, A9
    flowers, A6, A9
    germination, A33
    leaves, A5, A22, A42
    life cycles, A32–A33, A36
    mosses, A9
    nonvascular, A9
    photosynthesis, A5, A41, A49, A70
    pollination, A6, A32
    reproduction, A6–A7, A32–A33
    roots, A5
    seeds, A6, A32–A33
    spores, A7
    stems, A5
    transpiration, A42
    vascular, A8
pollination, A6, A32
pores (skin), D6
precipitation, B43, B55
prescription (drugs), D32–D34
producers, A50, A52, A70–A71
proteins, D56–D57, D62
protons, C5–C7
pulley, C103
pupa, A34

R

radial symmetry, A17
radiation (heat), C63
rain, B43, B48, B54–B56, R10
rain shadow, B48–B49
relative dating, B95
reproduction
    in animals, A20–A21
    in plants, A6–A7, A32–A33
resistors, C29
resources
    conservation of, B120, C67
    limited, A62

Index

rock(s)
  and fossils, B96–B97
  minerals, B100–B105
  relative dating, B95
  types, B92–B93
  weathering of, B65–B69
rock cycle, B94
rockets, space, B24
roots, A5
Rube Goldberg contraption, R8

S

sand, B65, B74, B111
satellites, B14, B24–B25
saturated fat, D55–D56
screw, C105
sediment, B74–B77
sedimentary rocks, B93–B96
seeds, A6, A32–A33
series circuit, C35, C37
short circuit, C39
side effects (chemical substances), D25–D27, D35
silt, B111
skin
  body temperature, D12–D14
  capillaries, D12–D13
  dermis, D6
  epidermis, D5
  function of, D7–D9, D12–D13
  fungi, D16
  hair follicles, D6
  integumentary system, D4, R13
  melanin, D5–D6
  parasites, D16
  pores, D6
  protecting your, D14–D17
  sweat glands, D6, D13
soil
  and plant growth, B111–B113
  composition, B110–B111
  composting, B121
  conservation, B120–B121
  destruction, B118–B119
  formation of, B108–B110
  horizons (layers), B117
  humus, B109–B110
  topsoil, B117–B119
  types, B110–B113
  water retention, B112
solar system
  definition, B5
  planets in, B5–B11
  satellites, B14
space exploration, B22–B27
space probes, B26
space shuttles, B27
species, A20–A21
speed, C93, R9
spore, A7
static electricity, C4, C8–C9, C24
steam, C70–C74
stems, A5
stimulants, D26
stratus clouds, B47
streak, B103
structural adaptation, A22–A23, A68–A69, A73
subsoil, B117
succession, A78–A79

Index

sun
   and living organisms, A5, A48, A52
   and tides, B18–B19
   gravitational pull, B5
   sunblock, D15, D17
sunlit zone (ocean), A72
sweat glands, D6, D13
switch, C27
symmetry, A17
symptom, D35

T

tadpole, A35
telescopes, B23, B24
Thales, C4
thawing, B66
thunderheads, B54
thunderstorms, B54
tides, B18–B19
tobacco, D42–D43
topsoil, B117–B119
tornado, B57
transpiration, A42, B45
twilight zone (ocean), A72

U

unsaturated fat, D55–D56

V

vaccines, D36–D37
vascular plants, A8
vertebrates, A14–A15
vitamins, D62–D63
volcano(es), B85
volume, C83

W

warm-blooded, A14
warm front, B54
water
   and erosion, B75–B77
   and soil, B112
   and weathering, B66–B67, B69
   and your body, D60–D61
   cycle, A42–A43, B42–B49
   ecosystems, A68–A73
   retention, B112
water vapor, B43
waves, B77
weather
   air masses, B52–B53
   fronts, B53–B55
   storms, B54–B57
weathering
   and water, B66–B67
   chemical, B65, B67
   definition, B65
   factors influencing, B68–B69
   freezing, B66
   physical, B65
   thawing, B66
wedge, C105
weight, C85
wheel and axle, C102
wind
   energy, B38, B39
   erosion, B74
   sources of, B36–B37

Z

zooplankton, A70

Credits

Photo Credits

Covers, Title Page, Unit Openers, NASA; **iv (t)**, ©Michael Sewell/Peter Arnold, Inc., **(b)**, ©Stefan Meyers/Animals Animals; **v**, ©Breck P. Kent/Animals Animals; **vi (tl)**. ©NASA/Science Photo Library/Photo Researchers, Inc., **(bl)**, Bernard Boutrit/Woodfin Camp & Associates; **vii (t)**, Grant Heilman from Grant Heilman Photography, **(b)**, Gary Randall/FPG International; **vii (tl)**, Mark C. Burnett/Photo Researchers, Inc., **(bl)**, Jim Corwin/Photo Researchers, Inc.; **ix (t)**, Alexander Lowry/Photo Researchers, Inc., **(b)**, ©David Stoecklein/The Stock Market; **x (tl)**, ©Jon Feingersh/The Stock Market, **(bl)**, ©Robert E. Daemmrich/Tony Stone Images, Inc.; **xi**, ©Nancy Sheehan/PhotoEdit; **xii, xiii, xiv, xv**, Matt Meadows; **A2-A3**, ©Michael Sewell/Peter Arnold, Inc.; **A6**, ©Joe McDonald/Earth Scenes; **A7 (tl)**, Matt Meadows, **(bl)**, ©Patti Murray/Earth Scenes, **(br)**, ©Richard Shiell/Earth Scenes; **A9 (t)**, ©Stephen J. Krasemann/DRK Photo, **(b)**, ©Darrell Gulin/DRK Photo; **A10, A11**, Studiohio; **A13 (tr)**, ©Oxford Scientific Films/Animals Animals, **(cl)**, ©James P. Rowan/DRK Photo, **(br)**, ©Zig Leszczynski/Animals Animals, **A14, (tl)**, ©E. R. Degginger/Animals Animals, **(bl)**, ©John Cancalosi/DRK Photo; **A15 (tl)**, ©Tom Brakefield/DRK Photo, **(tr)**, ©Nancy Rotenberg/Animals Animals, **(br)**, ©William Leonard/DRK Photo; **A16 (tl)**, ©Joe McDonald/Animals Animals, **(cl)**, ©Zig Leszczynski/Animals Animals, **(bl)**, ©Robert Maier/Animals Animals, **(br)**, ©C.C. Lockwood/Bruce Coleman, Inc.; **A17 (t)**, ©James P. Rowan/DRK Photo, **(c)**, ©M.C. Chamberlain/DRK Photo, **(b)**, ©Tim Rock/Animals Animals; **A18 (l)**, ©Betty K. Bruce/Animals Animals, **(r)**, ©Kenneth W. Fink/Bruce Coleman, Inc.; **A19 (tr)**, ©Jane Burton/Bruce Coleman, Inc. **(cr)**, ©William Leonard/DRK Photo, **(br)**, ©Lynn M. Stone/Bruce Coleman, Inc.; **A21 (tr)**, ©Stephen J. Krasemann/DRK Photo, **(cl)**, ©Ralph Reinhold/Animals Animals, **(b)**, ©Gerard Lacz/Animals Animals; **A22 (tr)**, ©Richard Shiell/Earth Scenes, **(cl)**, ©E.R. Degginger/Earth Scenes, **(br)**, ©Peter Parks/Oxford Scientific Films/Animals Animals; **A23 (tr)**, ©M.C. Chamberlain/DRK Photo, **(bl)** ©Stephen J. Krasemann/DRK Photo, **(bc)**, ©Jeremy Woodhouse/DRK Photo, **(br)**, ©E. R. Degginger/Earth Scenes; **A24**, ©Lewis Kemper/DRK Photo; **A25 (tr)**, ©Dimaggio/Kalish/The Stock Market, **(b)**, ©Zig Leszczynski/Animals Animals; **A26, A27**, © Matt Meadows; **A30-A31**, ©Stefan Meyers/Animals Animals; **A34 (tl, cl, bc)**, ©Dwight R. Kuhn/DRK Photo, **(br)**, ©D. Cavagnaro/DRK Photo; **A35 (tl)**, ©Breck P. Kent/Animals Animals, **(cr)**, ©Breck P. Kent/Animals Animals, **(b)**, ©Zig Leszczynski/Animals Animals; **A36**, ©Gerard Lacz/Animals Animals; **A37**, ©John Cancalosi/DRK Photo; **A38, A39, A41**, Matt Meadows; **A44**, ©D. Young-Wolff/PhotoEdit; **A45**, ©Dr. Nigel Smith/Earth Scenes; **A46, A47**, KS Studios; **A50 (tl)**, ©David Falconer/DRK Photo, **(ct)**, ©Jim Tuten/Earth Scenes; **(bl)**, ©Len Rue Jr./DRK Photo, **(br)**, ©John Visser/Bruce Colemann, Inc.; **A51 (tr)**, ©Stephen J. Krasemann/DRK Photo, **(cr)**,©E.R. Degginger/Earth Scenes, **(b)**, ©Michael Ederegger/DRK Photo; **A53 (t)**, ©Tom & Pat Leeson/DRK Photo, **(tcl)**, ©Bob Gurr/DRK Photo, **(tcr, cbl, cbr, cbc, cbcr)**, ©Stephen J. Krasemann/DRK Photo, **(cbcl)**, ©Darrell Gulin/DRK Photo, **(bc)**, ©Gary R. Zahm/DRK Photo, **(bl, bcr)**, ©D. Cavagnaro/DRK Photo, **(bcl)**, ©James P. Rowan/DRK Photo, **(br)**, ©Tom Bean/DRK Photo; **A54, A55**, Matt Meadows; **A58-A59**, ©Breck P. Kent/Animals Animals; **A61 (t)**, ©Bill Beatty/Animals Animals, **(cr)**, ©Stephen J. Krasemann/DRK Photo; **A63 (tr)**, ©Joe McDonald/DRK Photo, **(br)**, ©John Shaw/Bruce Coleman, Inc.; **A64**, ©R. & D. Aitkenhead/Animals Animals; A65, © Alex S. MacLean/Peter Arnold, Inc.; **A66, A67**, ©Matt Meadows; **A69 (tr)**, ©Bruce Watkins/Animals Animals, **(c)**, ©H. Taylor/Oxford Scientific Films/Animals Animals, **(bl)**, ©Doug Perrine/DRK Photo, **(br)**, ©Zig Leszczynski/Animals Animals; **A73 (tr)**, ©Prenzel Photo/Animals Animals, **(cl)**, **(bl)**, **(br)**, ©Norbert Wu/Peter Arnold, Inc.; **A74, A75**, Matt Meadows; **A77**, ©Andy Levin/Photo Researchers, Inc.; **A79 (tr)**, ©Michael Fredericks/Earth Scenes, **(cl)**, ©Ed Reschke/Peter Arnold, Inc., **(b)**, ©Jeffery Hutcherson/DRK Photo; **A80 (tl)**, ©Joseph VanWormer/Bruce Coleman, Inc., **(tr)**, ©Des & Len Bartlett/Bruce Coleman, Inc., **(bl)**, ©John Gerlach/DRK Photo; **A81 (t)**, ©Bill Wood/Bruce Coleman, Inc., **(cr)**, ©Eric Dragesco/Bruce Coleman, Inc., **(b)**, ©M.C. Chamberlain/DRK Photo; **A83**, Matt Meadows; **A87**, ©Lawrence Migdale/Tony Stone Images, Inc.; **B2-B3**, ©NASA/Science Photo Library/Photo Researchers, Inc.; **B6 (l)**, Photri, **(br)**, NASA/International Stock; **B7 (tl)**, NASA, **(br)**, Photri; **B8**, NASA/International Stock; **B9 (tr)**, Frank P. Rossotto, **(cl)**, Photri, **(br)**, NASA; **B11 (tr, cl)**, NASA/International Stock, **(b)**, NASA; **B12, B13**, Matt Meadows; **B15**, ©David Nunuk/Science Photo Library/Photo Researchers, Inc.; **B17**, ©Digital Image 1996/Corbis; **B18**, ©Jeff Greenberg/MRP/Photo Researchers, Inc.; **B20**, Matt Meadows; **B23**, ©NASA/Science Photo Library/Photo Researchers, Inc.; **B24**, Jonathan Blair/Corbis; **B25 (tr)**, ©David Ducros/Science Photo Library/Photo Researchers, Inc., **(bl)**, Corbis; **B26**, Corbis; **B27 (tl)**, AP/Wide World Photos, **(tr)**, Corbis/Bettmann; **B28, B29**, ©Matt Meadows; **B32-B33**, Bernard Boutrit/Woodfin Camp & Associates; **B35**, Jade Albert/FPG International; **B38 (l)**, Joe McDonald/Animals Animals, **(br)**, Travelpix/FPG International; **B39**, Peter Sterling/FPG International; **B40, B41**, Studiohio; **B43**, ©KS Studios; **B46**, Arthur Tilley/FPG International; **B47 (tl)**, Telegraph Colour Library/FPG International, **(br)**, ©Joyce Photographics/Photo Researchers, Inc.; **B50, B51**, ©KS Studios; **B53**, ©NOAA, colored by John Wells/Science Photo Library/Photo Researchers, Inc.; **B56**, Kal Muller/Woodfin Camp & Associates; **B57**, ©Howard Bluestein/Photo Researchers, Inc.; **B58, B59**, KS Studios; **B62-B63**, Grant Heilman/Grant Heilman Photography; **B65**, Bill Losh/FPG International; **B66 (tl)**, Walter H. Hodge/Peter Arnold, Inc., **(br)**, Grant Heilman/Grant Heilman Photography, **B67(bl)**, Toyohiro Yamada/FPG International, **(br)**, Dave Bartuff/Corbis; **B68**, ©SuperStock; **B69**, Douglas Mason/Woodfin Camp & Associates; **B70, B71**, Matt Meadows; **B73**, C. Allan Morgan/Peter Arnold, Inc.; **B74 (tl)**, ©W. Wisniewski/Okapia/Photo Researchers, Inc., **(bl)**, Telegraph Colour Library/FPG International; **B75**, ©Douglas Faulkner/Photo Researchers, Inc.; **B76 (tl)**, Barry L. Runk/Grant Heilman Photography, **(tr)**, ©Jim Steinberg/Photo Researchers, Inc.; **B77**, William Strode/Woodfin Camp & Associates; **B78**, Matt Meadows; **B82 (tl)**, ©Telegraphy Colour Library/FPG International, **(tr)**, Bill O'Connor/Peter Arnold, Inc., **(bl)**, AP/Wide World Photo, **(br)**, George Hall/Tony Stone Images, Inc.; **B85**, ©Hoa-Qui/Photo Researchers, Inc.; **B86, B87**, Matt Meadows; **B90-B91**, Gary Randall/FPG International; **B93 (tr)**, ©BioPhoto Associates/Photo Researchers, Inc., **(cr)**, ©Andrew J. Martinez/Photo Researchers, Inc., **(br)**, ©BioPhoto Associates/Photo Researchers, Inc., **(bl)**, Matt Meadows; **B95**, Jim Strawser/Grant Heilman Photography; **B96**, ©James L. Amos/Photo Researchers, Inc.; **B98, B99**, Matt Meadows; **B101 (cr)**, Runk/Schoenberger/Grant Heilman Photography, **(bl)**, Breck P. Kent/Earth Scenes; **B103 (tl)**, ©Charles D. Winters/Photo Researchers, Inc., **(tc)**, ©Aaron Haupt/Photo Researchers, Inc., **(tr)**, ©Ben Johnson/Science Photo Library/Photo Researchers, Inc., **(cb)**, Rosemary Weller/Tony Stone Images, **(bl)**, ©John Buitenkant/Photo Researchers, Inc., **(b)**, Matt Meadows; **B104 (tl)**, ©Martin Land/Science Photo Library/Photo Researchers, Inc., **(bl)**, ©Andrew M. Levine/Photo Researchers, Inc., **(br)**, First Image; **B105 (tr)** Julian Baum/Bruce Coleman, Inc., **(br)**, Robert E. Daemmrich/Tony Stone Images, Inc.; **B106, B107**, Brent Turner/BLT Productions/1991; **B109**, Matt Meadows; **B111**, Barry L. Runk/Grant Heilman Photography, Inc.; **B112**, Zig Leszczynski/Earth Scenes; **B113**, George Bernard/Earth Scenes; **B114, B115**, Brent Turner/BLT Productions; **B118**, Grant Heilman/Grant Heilman Photography; **B119**, ©Jacques Joangoux-Brazil State of Maranhao/Photo Researchers, Inc.; **B120**, Larry Lefever/Grant Heilman Photography; **B121**, Michael Newman/PhotoEdit; **B122, B123**, Matt Meadows; **B127**, KS Studios; **C2-C3**, ©Mark C. Burnett/Photo Researchers, Inc.; **C6**, KS Studios; **C8**, Mary Ann Evans; **C9**, KS Studios; **C10, C11**, Studiohio; **C13**, Tony Freemann/PhotoEdit; **C14**, KS Studios; **C15**, Matt Meadows; **C16 (cl)**, KS Studios, **(br)**, Anthony Meshkinyar/Tony Stone Images; **C17**, Rick Gayle/The Stock Market; **C18, C19**, Ken Karp; **C22-C23**, ©Jim Corwin/Photo Researchers, Inc.; **C25 (t)**, ©Matt Meadows, **(br)**, KS Studios; **C26 (tl)**, ©KS Studios **(b)**, ©Matt Meadows; **C27 (tl)**, Aaron Haupt, **(tr, cr, bl)**, KS Studios; **C28 (tl)**, ©David Young-Wolff/PhotoEdit, **(br)**, KS Studios; **C29**, First Image; **C31**, Matt Meadows; **C32, C33**, KS Studios; **C35**, Matt Meadows; **C37 (tl)**, Frank Saragnese/FPG International, **(br)**, Richard Gaul/FPG International; **C38 (cr)**, Aaron Haupt, **(bl)**, KS Studios; **C41**, Ken Karp; **C43**, KS Studios; **C44**, Matt Meadows; **C45**, E.R. Degginger/Earth Scenes; **C46 (cl)**, Matt Meadows, **(br)**, ©Tony Freeman/PhotoEdit; **C47**, ©David R. Frazier/Photo Researchers, Inc.; **C48**, Ken Karp; **C52-C53**, ©Alexander Lowry/Photo Researchers, Inc.; **C54**, First Image; **C56**, KS Studios; **C57**, Richard Hamilton Smith/Corbis; **C58 (tl)**, ©Theodore Clutter/Photo Researchers, Inc., **(tr)**, ©Calvin Larsen/Photo Researchers, Inc.; **C59**, ©Paolo Koch/Photo Researchers, Inc.; **C60, C61**, KS Studios; **C63 (tr)**, KS Studios, **(br)**, ©Catherine Ursillo/Photo Researchers, Inc.; **C64**, KS Studios; **C67**, Yoav Levy/Phototake; **C68-C69**, Matt Meadows; **C71**, J.R. Holland/Stock Boston; **C73**, KS Studios; **C74**, ©David Young-Wolff/PhotoEdit; **C75**, ©Kees van den Berg/Photo Researchers, Inc.; **C76, C77**, ©Matt Meadows; **C80-C81**, ©David Stoecklein/The Stock Market; **C82, C83, C84**, KS Studios; **C86 (cl)**, KS Studios, **(br)**, Matt Meadows; **C87, C88, C89**, KS Studios; **C91**, Allsport/Vandystadt; **C92 (tl)**, ©Pete Saloutos/The Stock Market, **(br)**, ©Telegraph Colour Library/FPG International; **C93**, ©DiMaggio/Kalish/The Stock Market; **C94**, ©David Young-Wolff/PhotoEdit; **C95 (tr)**, ©J. Taposchaner/FPG International, **(br)**, ©Stephen Dalton/Photo Researchers, Inc.; **C96, C97**, KS Studios; **C99 (tr)**, Jim Cummins/FPG International, **(br)**, Matt Meadows; **C100**, KS Studios; **C101 (tl)**, KS Studios, **(br)**, ©Peter Beck/The Stock Market; **C102 (tl)**, KS Studios, **(br)**, Matt Meadows; **C103**, ©Blair Seitz/Photo Researchers, Inc.; **C104**, Tony Freeman/PhotoEdit; **C105, C106**, KS Studios; **C111**, Matt Meadows; **D2-D3**, ©Jon Feingersh/The Stock Market; **D5 (tr)**, ©Professors P.M. Motta, K.R. Porter & P.M. Andrews/Science Photo Library/Photo Researchers, Inc., **(br)**, ©Stephen Simpson/FPG International; **D8**, ©Tony Freeman/PhotoEdit; **D9**, ©David Young-Wolff/PhotoEdit; **D10, D11**, KS Studios; **D13**, ©David Young-Wolff/PhotoEdit; **D14 (tl)**, ©David Woods/The Stock Market, **(br)**, ©Barbara Stitzer/PhotoEdit; **D15 (tr)**, ©David Harry Stewart/Tony Stone Images, **(br)**, ©Roy Morsch/The Stock Market; **D16 (tl)**, ©Oliver Meckes/Photo Researchers, Inc., **(tr)**, Visuals Unlimited/©John D. Cunningham, **(bl)**, ©Oliver Meckes/Photo Researchers, Inc.; **D17**, ©Jacob Taposchaner/FPG International; **D19**, Matt Meadows; **D22-D23**, ©Robert E. Daemmrich/Tony Stone Images, Inc.; **D25**, ©Steve Smith/FPG International; **D26**, KS Studios; **D27**, ©Cristoph Wilhelm/FPG International; **D28**, Visuals Unlimited/©M. Eichelberger; **D29**, ©Charles D. Winters/Photo Researchers, Inc.; **D30**, Matt Meadows; **D33**, ©Michelle Bridwell/PhotoEdit; **D34**, ©David Young-Wolff/PhotoEdit; **D35**, ©Steve Smith/FPG International; **D36**, ©Amy C. Etra/PhotoEdit; **D38, D39**, ©First Image; **D42**, ©Frank Saragnese/FPG International; **D43 (tr)**, ©Custom Medical Stock Photo, **(br)**, ©Science Photo Library/Custom Medical Stock Photo; **D46**, Matt Meadows; **D50-D51**, ©Nancy Sheehan/PhotoEdit; **D53**, KS Studios; **D54**, First Image; **D55**, KS Studios; **D56**, ©Michael Newman/PhotoEdit; **D57**, ©Rudi Von Briel/PhotoEdit; **D58**, KS Studios; **D59**, Matt Meadows; **D61**, ©David Young-Wolff/PhotoEdit; **D62**, First Image; **D64**, KS Studios; **D65**, ©Myrleen Ferguson/PhotoEdit; **D66**, Matt Meadows; **D67**, First Image; **D71**, KS Studios; **R2 (cl)**, John Lemker/Earth Scenes; **R2-R3**, ©S. Nielsen/DRK Photo; **R3 (cr)**, George Godfrey/Earth Scenes, **(br)**, Michael Fogden/Earth Scenes; **R6 (bl)**, NASA/Science Photo Library/Photo Researchers, Inc., **(br)** Brent Turner/BLT Productions; **R6-R7**, ©John Bova/Photo Researchers, Inc.; **R7 (tr)**, NASA/Science Photo Library/Photo Researchers, Inc.; **R9 (tr)**, Terry Murphy/Animals Animals, **(cr)**, John Lemker/Animals Animals, **(bl)**, Perry D. Slocum/Animals Animals, **(br)**, Robert Lubeck/Animals Animals; **R10 (tl)**, Telegraph Colour Library/FPG International, **(cr)**, Corbis/Douglas Peebles, **(bl)**, Corbis/Andrew J. G. Bell; Eye Ubiquitous, **R11 (tr)**, E.R. Degginger/Earth Scenes, **(cl)**, Michael Fredericks, Jr. /Earth Scenes; **R12**, Gabe Palmer/The Stock Market.

Art Credits

A5 A6 A41 A49 A62 A70 A72 Sandra McMahon/McMahon Medical Art; **B5 B10 B15 B16 B19** Rolin Graphics Inc.; **B54 B55** SRA; **B94** Rolin Graphics Inc.; **C30 C55 C65 C66 C84-85 C103 C104 C105 D5 D6 D7 D18 D19 D33 D37 D41 D45** Precision Graphics; **D63** Chris Higgins/PP/FA.